E. A Kirby

A Formulary of Selected Remedies

With Therapeutic Annotations, Adapted to the Requirements. Second Edition

E. A Kirby

A Formulary of Selected Remedies
With Therapeutic Annotations, Adapted to the Requirements. Second Edition

ISBN/EAN: 9783337014711

Printed in Europe, USA, Canada, Australia, Japan

Cover: Foto ©berggeist007 / pixelio.de

More available books at **www.hansebooks.com**

A FORMULARY

OF

SELECTED REMEDIES

WITH THERAPEUTIC ANNOTATIONS,

ADAPTED TO THE

REQUIREMENTS OF GENERAL PRACTICE,
HOSPITALS, DISPENSARIES, PARISH
INFIRMARIES, LUNATIC ASYLUMS, AND OTHER
PUBLIC INSTITUTIONS.

WITH INDEX OF DISEASES AND REMEDIES,
DIET TABLES, ETC.

BY

E. A. KIRBY, M.D., M.R.C.S.ENG.,

Late Physician to the City Dispensary.

SECOND EDITION.

LONDON:
H. K. LEWIS, 136, GOWER STREET, W.C.

1874.

Butler & Tanner,
The Selwood Printing Works,
Frome, and London.

CONTENTS.

PREFACE.

The object of this little work is to aid the busy practitioner in the selection and administration of the medicines he prescribes, and to help him in *dispensing*— an occupation which forms a necessary, often an irksome, but seldom an agreeable, part of his daily occupation.

Formularies or Pharmacopœias more or less complete are employed in all the London and Provincial Hospitals, Dispensaries, and Infirmaries. They are compiled with reference to the special requirements of the Institutions in which they are employed, and are designed to economise not only materials but also the time and labour of the medical officers and dispensers, and thereby to facilitate materially the work of the Institution. Stock Medicines, *e.g.*, Concentrated Mixtures and Solutions and Pills, prepared according to adopted formulæ, are commonly employed with the same object in private practice. But as far as I am aware there has been no attempt to compile a Formulary adapted to the modern requirements of General Practice, with a view to supersede as far as possible extemporaneous dispensing, and to embrace all the improvements and elegancies of modern pharmacy. The present work has been undertaken with a view to supply such a Formulary. The practitioner will select those adapted to his individual requirements, and these he may obtain ready for use, and so save himself the trouble of preparation. The formulæ are all reliable, none having been admitted unless sanctioned by good authority or tested by my own experience,—they are sufficiently comprehensive to supply a medicine for all ordinary cases, and as most of the remedies have the advantage of extreme portability, those most commonly required can without inconvenience accompany him on his rounds, and be at once dispensed—an obvious advantage to himself and his patients. This is more practicable now than heretofore, because the medicinal requirements of modern practice are encompassed within much narrower limits, and the practitioner may, with much advantage, dispense with the cumbrous and complicated appliances which were formerly necessary to the pharmacy of general practice. I solicit my *confrères* to apprise me of any inaccuracies or omissions that may come under their notice, and I shall be glad to receive any suggestions tending to promote the usefulness of future editions.

E. A. KIRBY.

26, *Gordon Square.*

INTRODUCTION:

THE FOLLOWING FORMS are selected with the special object of reducing the labour of dispensing to a minimum, without sacrificing either efficiency or utility.

FOR INTERNAL ADMINISTRATION.

1. *Fluids.*
   - Drops.
   - Mixtures.
   - Potions, or Drinks.
   - Syrups.
   - Elixirs.

2. *Solids.*
   - Pills, Pilules, and Granules.
   - Lozenges and Glycecols.

3. *Vapours or Inhalations.*

FOR EXTERNAL OR TOPICAL ADMINISTRATION.

- Baths.
- Poultices.
- Fomentations.
- Paints.
- Injections.
- Liniments.
- Lotions.
- Ointments.
- Caustic Pastes.
- Pessaries, Suppositories, and Bougies.
- Collyria.
- Gargles, Lozenges, and Glycecols.

GUTTÆ(DROPS).—When, for the sake of portability or other-wise, fluid concentrated preparations are needed, a definite number of DROPS may be ordered to be taken in water; but for general purposes it is better to adopt the teaspoonful dose in a wineglassful of water, and to regulate the proportions of the active ingredients accordingly. This method is both prac-ticable and convenient, and is especially applicable to the administration of the dilute mineral acids, tinctures, *e.g.*, tincture aconite, tincture veratrum viride, etc., expressed juices, elixirs, and some of the liquors. The dose may be de-termined with precision, as vials graduated for 4, 6, 8, and 12 teaspoons may be obtained at the medical glass warehouses.

MISTURÆ (MIXTURES).—For the administration of some medicines this form is indispensable, but that described above may, with advantage, be substituted for many which are con-stantly prescribed in *mixtures*, more from the force of habit and custom than from necessity. The drugs which necessarily are given in this form are those which are sparingly soluble, or require to be freely diluted. The Salts of Magnesia, Soda, and Potash, are of this kind, and Infusions, Decoctions, and other such like medicaments are to be given in this form. Mixtures are largely employed in Hospital Practice, partly, I believe, to humour a popular prejudice in their favour. Out-patients especially expect their medicine in a fluid form. It satisfies the patient. A large bottle of medicine they think is some-thing costly, obtained cheaply, and rewards them for the time and labour bestowed in the getting. More costly medi-cine in a less bulky form would not find so much favour in their eyes. The same silly prejudice exists among parish and club patients to the serious inconvenience of the practitioner, who is obliged to study the man as well as the patient. Many practitioners use simple concentrated mixtures, while, for the sake of convenience and economy, they depend chiefly upon the pilular form for the essential part of their medicinal treatment.

The Hospital Pharmacopœias give numerous formulæ for mixtures, and considerable confusion is occasioned by the same names representing different medicines at different hospitals. In some instances the variations of materials or quantities are very trifling, in others they are very material. At page 51, will be found the formulæ of those most in common use at the hospitals. They are not identical with any, while they are typical of all. These formulæ might be largely extended, but those given are sufficient for general purposes.

PILULÆ.—It is convenient to divide this form into pills, pilules, and granules ; *pilule* being used as a diminutive to denote a *small* pill ; granules, or grains, being pilules not exceeding a grain in weight. Some parents who would object to pills, willingly give their children granules.

Of all forms of medicines this is now the most convenient to the practitioner, and generally the most popular with his patients. The manufacture of medicines in this form has received considerable attention, and is a distinct *branch* of pharmacy. Pill making has always been regarded by the practitioner as the most unpleasant part of his necessary dispensing. Now they can be obtained *ready made*, and better made than it is possible to prepare them extemporaneously ; hence they offer to him the most convenient mode of supplying his patients with remedies.

GLYCECOLS, a new form which I have devised for administering medicines, are fully described at page 64. They supply an effective and elegant, as well as convenient, mode of applying certain remedies locally to the throat, often superseding the necessity of using gargles, and are an attractive medium for administering powders to children.

SPRAY.—Medicated fluids, atomised by means of spray tubes, may be most advantageously applied in affections of the lungs, trachea, mouth, and fauces. In croup, diphtheria, and inflammations of the air passages, they are employed with great advantage. Drugs suitable for this form of administration are named at page 106.

HYPODERMIC INJECTION.—Solutions of Morphia, Atropine, and Aconitine are employed in this form with advantage in cases of advanced Cancer, Tetanus, Rheumatism, Neuralgia, etc. Formulæ for those commonly employed are given at page 112.

MEDICINES FOR CHILDREN.

It is exceedingly desirable that medicines for children should be prepared palatable, and, if possible, tasteless. To this end I have endeavoured to supply formulæ adapted to the treatment of all the diseases of childhood.

The Pilular and Glycecol forms are the best for this purpose. In the pill form they may be prepared so small and attractive looking for infants and young children that the nurse or parent can administer them without the knowledge of the patient, if very young, and readily with the consent of those who are older (as they can be made quite tasteless), without resorting to arts of deception and false assurances so commonly practised and so quickly detected. Induced to taste and try, a child's confidence is soon obtained, and opposition ceases when the medicine is found to be tasteless or tolerably palatable.

When powders are necessary, the Glycecol offers an admirable medium for administering them to children, especially to infants and those too young to be educated to swallow granules. The reader will find a number of formulæ prepared in this manner, in all of which the taste of the medicine is well concealed. It is both cruel and unnecessary to give nauseous medicines to children. Under the head of *Mistura* also several formulæ will be found, which have been specially devised to meet the ordinary requirements of diseases common to childhood.

REMARKS ON A READY METHOD OF DISPENSING MEDICINES.

As long ago as 1866 I suggested a method of dispensing, by means of *Portable Miniature Dispensaries and Ready Remedies.* It consisted essentially in this—

1st. That the practitioner should adopt a Formulary of Remedies adapted to the requirements of his own practice, and keep these medicines in a state of *readiness to administer*, rather than in a *crude state*, which necessitates pharmaceutical manipulation, involving time and labour, before they can be dispensed to his patients ; and with this object the present Formulary was in part compiled.

2nd. By means of a Miniature Dispensary, which is simply a box or case (divided into many compartments, each being fitted with a suitable tube or bottle), in which is carried a sufficiently comprehensive selection of the adopted formulæ to answer all the ordinary indications for medicinal treatment.

This method has been found to greatly conduce to the convenience of the practitioner and the comfort and benefit of his patients, and the present work is offered as an aid to its more general adoption ; in it the formulæ are more numerous, and many of them have been selected with special reference to this method of administration.

Among the many advantages which the adoption of *this* method offers, the most important is that it enables the practitioner to supply most of the medicinal requirements of his patients at the time they are seen, whether it be by the bedside far off, or in his own consulting room at home. And only those who are actively engaged in general practice can appreciate how great a desideratum this is, and what relief it affords him in his work. To return home from a long round, with all, or nearly all his patients supplied with medicine, and to be spared the toil of recalling the particular requirements of each, and the labour of dispensing for all, is a boon which is worth taking some pains to secure. Moreover, patients invariably appreciate very highly the prompt manner in which these Ready Remedies are supplied. However opinions may differ as to the *power of drugs to cure disease*, there is no doubt as to their power to alleviate pain and suffering, and the value of the visit is greatly enhanced when *present distress* is relieved at once. Practitioners often visit patients miles from their home, taking no medicine with them, so that the patient after waiting, it may be, many hours for the doctor, has to submit to the further delay and inconvenience of sending to his surgery for the remedy, however urgently needed—it may be a styptic to arrest hæmorrhage which threatens life, or a narcotic to relieve intolerable suffering.

This ready method necessitates, of course, some change in the mode of prescribing, which at the commencement demands, to a certain extent, some thought and painstaking, but the result amply rewards the trouble taken.

PILL COATING.

The value of the pilular form of administering medicines has been immensely increased, and its general applicability extended by the process of coating or enamelling, now so popular. Properly done it effectually protects the pills from atmospheric influence, and preserves the medicament in a *fresh and moderately moist condition,* ensuring its activity for a considerable length of time. Moreover it conceals the taste of the drug and gives the pill a pleasant appearance in place of a repulsive one, thereby making them acceptable to those who otherwise could not be induced to take them. The materials employed, French Chalk and Gums in aqueous or alcoholic solutions, are perfectly harmless, and cannot possibly exert a deleterious effect of any kind. The secret of successful coating is to be sought in the manner in which these are put on, in other words in the process of manufacture. It is a very simple matter indeed to moisten a pill with mucilage, roll it about in French Chalk until it assumes a white pearly appearance, and to call it coated. Such a coating, which is used by some makers, is not permanent, but quickly perishable ; its inexpensiveness, however, is its chief attraction to some, and its demerits are covered by a confident assurance of its *immediate solubility*. Every practitioner knows full well that there are some conditions in which pills, like other substances, meat, fruit, etc., are hurried through the bowels unchanged, and this is an accident dependent more on the condition of the patient than on the substance ingested. The most absurd tests of solubility are constantly put forth, and are calculated to mislead the judgment of the purchaser. It is quite possible for a coating to wash off in cold water in a few moments, and yet the pill itself to be quite inoperative. The conditions which govern solubility in the stomach are *not exactly* comparable with those of cold water in a tumbler. Temperature and the solvent powers of the secretions have to be taken *slightly into account*, and in this matter the proof of the value of a coated pill must be judged like the cook's pudding. The coating may be made a cloak for defective manufacture. Much more important than the coating is the quality of the pill itself; it should be tested by its degree of hardness, its uniform consistence, and, if necessary, by chemical and microscopical examination. The ancient process of coating with GOLD OR SILVER LEAF has recently been revived. It is very elegant, but costly, and not so effective a preservative as other materials.

SUGAR COATING.—This process, which is similar to that used in comfit making, necessarily enlarges the pills, but the chief objection is that they must be dried and exposed to considerable heat, thereby impairing the quality of the ingredients, and rendering the pill insoluble by depriving it of moisture.

CLASSIFICATION OF FORMULÆ.

Antacids, Antilithics, and Absorbents.—F. 50, 109, 157,* 159, 188*; Mistura Alkalina (Potash) c. Gentianâ; Mistura Alkalina (Soda) c. Calumbâ; Mistura Alkalina Aromatica; Mistura Carminativa Antacid.*

Alteratives and Resolvents.—F. 1 to 8, 10 to 14, 16, 17, 24, 27* to 30, 39, 92, 104, 124, 127, 129, 133, 148, 149, 150, 155, 163, 175, 188; Mist. Hydrargyri Co. et Sarsæ.; Mist. Potassii Iodidi Co.; Mist. Potassii Bromidi Co.

Anthelmintics.—F. 42, 43, 44, 51.

Antispasmodics.—F. 18, 26, 31* to 34, 40, 93, 105, 106, 120, 136, 152, 153, 158, 182, 185.

Astringents.—F. 25, 35 to 38, 40, 130, 138; Mist. Acid. Sulph. Arom.; Mist. Acid. Phosphorici c. Ferro; Mist. Astringens c. Hæmatoxyli; Mist. Astringens;* Syr. Krameriæ;* Pulv. Astringens.

Cathartics, Cholagogues and Hydragogues.—F. 19, 22, 23, 41* to 43,* 51* to 56, 60 to 63,* 65, 66,* 67,* 87, 116, 118, 119, 126, 131, 137,* 141, 142, 147, 154, 164, 164A, 171, 174, 187; Mist. Magnes. Sulph. Acid.; Mist. Magnes. Sulph. Alkalina; Mist. Mag. Sulph. c. Rosâ; Mist. Carminativa Aperiens;* Syr. Rhei Aromat.* (Spiced Syr. Rhubarb).

Diaphoretics and Salines.—F. 98 to 100, 102,* 110, 112, 133, 148, 149, 165, 167; Mist. Ammoniæ Acet.; Mist. Ammoniæ Acet. Comp.; Mist. Ammoniæ Effervescens; Pulv. Salinæ Effervescens.

Diuretics.—F. 9, 101, 104, 162, 181; Mist. Copaibæ Co.; Mist. Diuretica; Mist. Salinæ.; Tr. Colchici Eth.; Tr. Guaiaci Eth.

Digestants.—F. 45, 48, 49, 50, 76, 151, 156, 172, 180. Elixir Pepsinæ; Elixir Bismuthi; Mist. Alkalina Aromat. c. Rheo; Mist. Acid. Nitrohydrochloric.

---

* Formulæ marked thus (*) are medicines for children. See also GLYCECOLS, page 64.

Emmenagogues.—F. 68, 69, 87, 105, 106, 107, 168.

Expectorants, Sedative and Stimulating.—F. 96, 111, 117, 123, 166, 183 ; Syr. Ipecacuanhæ ;* Syr. Senegæ. ; Mist. Ammoniaci Ipecac. et Lobeliæ ; Mist. Ipecacuanhæ ;* Mist. Cascarillæ Co. ; Mist. Senegæ Co.

Narcotics, Anodynes, Hypnotics, and Soporifics.—F. 35, 36, 88, 91,* 95, 117, 123, 130, 139, 140, 153, 176, 177, 186 ; Tinct. Opii Etherea ; Tinct. Opii c. Chloroformi.

Sedatives and Depressants.—F. 89, 90, 91,* 92, 114, 115, 133 to 135, 167 ; (Arterial) Tinct. Veratri Viridis ; U. S.—Antimony ; Aconite.

Stimulants, Special.—F. 57, 58, 93, 125, 146, 189 to 200. See also ALTERATIVES.

Stimulants, Cardiac.—Alcohol ; Ammonia (Mist. Cinchonæ Ammon.) ; Turpentine Enemata ; Belladonna and Digitalis, preparations of.

Tonics and Antiperiodics.—F. 12, 13, 20, 21, 44*, 47, 64, 70 to 86,* 113, 120 to 122, 127, 128, 132, 143 to 145, 150, 157, 160, 161, 168 to 170, 172, 173, 178, 179, 184, 189 to 200. Elixir Bark ;* Elixir Bark and Iron ;* Elixir Pyrophosphate Iron ;* Elixir Gentian and Iron ; Mist. Acidi Nitrohydrochlorici ; Mist. Acid. Phosph. c. Ferro ; Mist. Acid. Nitrohydrochlor. c. Ferro et Strychniâ ; Mist. Chiratæ Co. ; Mist. Cinchonæ Acida ; Mist. Cinchonæ Ammon. et Chloroformi ; Mist. Cinchoniæ.

FORMULÆ.

PART I.—INTERNAL REMEDIES.

PILULÆ.

vomiting and pain.

(3.)
Hydrarg. Subchlor. c. Opio.

℞ Hyd. Subchlor., gr. ij.; Pulv. Opii., gr. ½. M. ft. pil.
To produce ptyalism. One pill every three or four hours.

B

Emmenagogues.—F. 68, 69, 87, 105, 106, 107, 168.

Expectorants, Sedative and Stimulating.—F. 96, 111, 117, 123, 166, 183; Syr. Ipecacuanhæ;* Syr. Senegæ.; Mist. Ammoniaci Ipecac. et Lobeliæ; Mist. Ipecacuanhæ;* Mist. Cascarillæ Co.; Mist. Senegæ Co.

Narcotics, Anodynes, Hypnotics, and Soporifics.—F. 35, 36, 88, 91,* 95, 117, 123, 130, 139, 140, 153, 176, 177, 186; Tinct. Opii Etherea; Tinct. Opii c. Chloroformi.

Sedatives and Depressants.—F. 89, 90, 91,* 92, 114, 115, 133 to 135, 167; (Arterial) Tinct. Veratri Viridis; U. S.—Antimony; Aconite.

Stimulants, Special.—F. 57, 58, 92, 125, 146, 180 to 200.   See

FORMULÆ.

PART I.—INTERNAL REMEDIES.

PILULÆ.

(1.)
Hydrarg. Subchlor. c. Opio.

℞ Hyd. Subchlor., gr. iij.; Pulv. Opii, gr. j.   M. ft. pil.

CALOMEL, although not so largely employed as formerly, is still held to be a most valuable antiphlogistic in inflammatory and febrile affections—a sheet anchor in the treatment of inflammation of membranes, especially *serous* membranes, *peritonitis, pleurisy, pericarditis*, also in inflammation of the tissue of the eye, and *iritis*. OPIUM promotes its antiphlogistic powers, and prevents its acting on the bowels.   However this combination be employed, whether as a sedative or to produce ptyalism, one or other of the following six formulæ will conveniently meet the requirements of most cases.

See also Nos. 2 to 6.

(2.)
Hydrarg. Subchlor. c. Opio.

℞ Hyd. Subchlor., gr. ij.; Pulv. Opii., gr. j.   M. ft. pil.

A useful remedy at the commencement of an attack of cholera and diarrhœa, one or two doses being sufficient in most cases to relieve vomiting and pain.

(3.)
Hydrarg. Subchlor. c. Opio.

℞ Hyd. Subchlor., gr. ij.; Pulv. Opii., gr. ½.   M. ft. pil.

To produce ptyalism.   One pill every three or four hours.

B

(4.)

Hydrarg. Subchlor. c. Opio.

℞ Hyd. Subchlor., gr. ij. ; Pulv. Opii, gr. ¼. M. ft. pil.

To be preferred when full doses of opium are *contra-indicated*.

(5.)

Hydrarg. Subchlor. c. Opio.

℞ Hyd. Subchlor., gr. j. ; Pulv. Opii, gr. ½. M. ft. pil.

To produce rapid ptyalism, with very little disturbance of the system generally. One every two hours, for 12 or 24 hours, watching the effect.

(6.)

Hydrarg. Subchlor. c. Opio.

℞ Hyd. Subchlor., gr. j. ; Pulv. Opii, gr. ¼. M. ft. pil.

The same as above. One may be taken every hour until salivation is produced.

> *See Calomel and Antimony* . . . F. 133
>   ,,   ,,   *with Opium* . F. 134
>   ,,   *and Dover's Powder* . . F. 110
>   ,,   *and James's Powder* . . F. 112

The above are all useful remedies in the treatment of acute inflammations of the sthenic type.

> *See also Aconite and Opium, F.* 90.

(7.)

Hyd. c. Cretâ et Hyoscyami.

℞ Hyd. c. Cretâ, gr. iij. ; Ext. Hyoscyami, gr. ij. M. ft. pil.

A very useful form for bringing the system mildly under the influence of Mercury, *in syphilitic affections*, etc.

> See Grey Powder with Dover's Powder . F. 92 and 149.
> Blue Pill and Opium . . ` . . F. 29 and 148.

(8.)
Hydrarg. Perchlor.

℞ Hydrarg. Perchlor., Ammon. Chlorid., aa. gr. xij. Ft. gran. 240.

PILULÆ SUBLIMAT. CORROSIV. (*Dzondi.*) Prescribed with excellent effect in *syphilitic secondary affections, chronic skin diseases,* and in all cases where the *alterative* effect of mercury is desired. Not so likely to produce ptyalism as other mercurial preparations.

Dose, one three times a day, increased gradually until six or eight pills are taken daily.

See Corrosive Sublimate with Quinine, F. 146.

(9.)
Pil. Hydrarg., Scillæ, et Digitalis.

℞ Pil. Hydrarg., gr. iij.; Pulv. Digitalis, gr. ½; Pulv. Scillæ, gr. jss. M. ft. pil.

A useful alterative and diuretic in hepatic and cardiac dropsy.

See Digitalis and Squills	F. 101
,, ,, *and Calomel* . .	F. 104

(10.)
Podophylli et Ipecac.

℞ Podophylli Res., gr. ½; Pulv. Ipecac., gr. ½; Ext. Hyoscyami, gr. ij.; Pulv. Capsici, gr. ½. M. ft. pil.

PODOPHYLLUM PELTATUM is a powerful cholagogue and alterative. It is given with excellent effect in suppression or partial suppression of the secretion of bile, is useful in *hepatic enlargements, dropsy,* and other disorders of persons who have long resided in hot climates. The above formula will be found most efficient when its alterative action is desired.

"As a simple alterative it is as valuable as mercury, without possessing any injurious qualities."—DR. TANNER.

Dose—One, twice or thrice a day. An occasional, dose ot *Friedricshalle Water* taken fasting (during a course of Podophyllin) produces copious bilious evacuations.

See Podophyllin with Compound Colocynth .	. F. 60
,, ,, *Compound Rhubarb* .	. F. 119
,, ,, *Rhubarb and Henbane*	. F. 53

Also Podophyllin Granules (lactinated) containing gr. $\frac{1}{60}$. Alterative for infantile constipation.

(11.)
Pot. Iodid. et Colchici.

℞ Pot. Iodid., gr. ijss.; P. Sem. Colchici, gr. ij.; Ext. Aconiti, B. P., gr. ½. M. ft. pil.

This and the two following formulæ are recommended as a con-venient and pleasant mode of administering Iodide of Potassium. Efficacious in the treatment of chronic *gout* and *rheumatism*, especially when complicated with *constitutional syphilis, syphilitic iritis*, and *retinitis.*

See F. 30.

(12.)
Pot. Iodid. c. Quinâ.

℞ Pot. Iodid, gr. ijss.; Quinæ Sulph., gr. j. M. ft. pil.

Useful in removing "pains," vaguely described as "rheumatics flying about," in cases of debility where Colchicum is contra-indicated. For sudden attacks of muscular rheumatism, for pleuro-dynia, etc., this acts almost as a specific.

(13.)
Pot. Iodid. et Ferri Cit. c. Quinâ.

℞ Pot. Iodid., gr. ij.; Ferri Cit. c. Quinâ, gr. ijss. M. ft. pil. *

As above, and in secondary and tertiary syphilis in weak and anæmic subjects.

Dose—One or two three times a day.

Note—IODIDE OF POTASSIUM (*simple*) is conveniently prescribed in the form of Pearl Coated Pills, each containing either three or five grains. Patients who will not take it in solution, readily submit to a course of the Iodide when prescribed in this manner. These pills have the further advantage of portability, and they will keep in any climate. See F. 155.

(14.)
Hyd. Iodid. Vir. c. Hyoscyamo.

℞ Hyd. Iodid. Vir., gr. j.; Ext. Hyoscy., gr. ij. M. ft. pil. *

An elegant and efficacious mode of administering this useful alterative. It is given with the best effect in pustular and tuber-cular diseases of the skin and in constitutional syphilis.

Dose—One night and morning.

See Iodide of Mercury with Soda, F. 124. Alterative for children in skin diseases.

---

* Half this strength it is a useful remedy for the pustular keratitis of children.

Creasoti.

(15.)

℞ Creasote, min. xx. ; P. Aromat., gr. lxxx.   M. ft. pil. 20.

Creasote is employed to check nausea and retching occurring in various diseases, also in sea-sickness, pregnancy, etc.   Sometimes creasote will give great relief by arresting the vomiting caused by malignant disease of the stomach.   In these pills it is easily taken. It has been given with good effect to check profuse expectoration in chronic bronchitis.

Dose—One pill three or four times a day.

Small doses, *often repeated*, act more efficaciously than large doses. This remark applies to a large number of medicines.   We are too much bound by habit and custom in this respect.   There is much to be said in favour of the gradual introduction of active medicines into the system.

(16.)

Hyd. Iodid. et Morphiæ.

℞ Hyd. Iodid. Rub., gr. jss.; Morphiæ Hydroch., gr. j.   M. ft. pil. 12.

An anodyne alterative, useful in ulcerous and tubercular diseases of the skin, especially in syphilitic constitutions.

Dose—One night and morning.

See also No. 80.

(17.)

Hyd. Iodid. et Arsenici.

℞ Hydrarg. Iodid. Rub., gr. j. ; Arsenici Iodid., gr. j. ; Potass. Iodid., gr. xx.   M. ft. pil. 20.

Intended as a convenient substitute for Donovan's Solution.   It has been largely prescribed, and is a very favourite remedy with many practitioners.   It is given with excellent effect in psoriasis, and in obstinate squamous and ulcerous diseases of the skin, as well as in cachectic cases, in which both iodine and arsenic in minute doses are very successful.

Dose—One three times a day, after food.

(18.)
Ammon. Bromid. et Valer.

℞ Ammon. Bromid., gr. iij. ; Ext. Valerianæ, gr. j. M. ft. pil.

Useful in functional diseases of the nervous system, hysteria, etc. It often tranquilizes the pulse, and induces sleep in restless cases of nervous excitement. Also a valuable absorbent in glandular enlargements. A good sedative in hooping cough.

Dose—One three times a day. To produce sleep, three at bedtime.

(19.)
Aloin et Podophylli.

℞ Aloin, gr. j. ; Pódophylli Res., gr. ½ ; Olei Zingib., m ⅛. M. ft. pil.

An American remedy for constipation, said to act well when taken with dinner.

Dose—One at dinner time.

(20.)
Ipecac. et Quinæ.

℞ Quinæ Sulph., gr. viij. ; Pulv. Ipecacuanhæ, gr. xxiv. ; P. Ipecac. c. Opio, gr. xxx. M. ft. pil. 18.

This combination is useful in sub-acute dysentery and in affections of the mucous surfaces. In chronic bronchitis when debility is prominent and the cough severe it is also valuable.

(21.)
Quinæ et Belladon.

℞ Quinæ Sulph., gr. ij. ; Ext. Belladonnæ, gr. ⅓ ; Ext. Opii, gr. ½ ; Ext. Hyoscyami, gr. ij. M. ft. pil.

A useful "*Pain Killer*" in neuralgia and carcinoma, as well as a sedative in pruritis of the vulva.

Dose—One every six or eight hours.

N.B.—Extract of Opium is fully one third stronger than crude Opium.

See also No. 71, and note to 95.

(22.)
Cal., Colchici, et Aloes.

℞ Hydr. Subchlor, Ext. Colchici. Acet., Ext. Aloes.
Barb., Pulv. Ipecac., aa. gr. j. M. ft. pil.
See Gout Pills, F. 116 *and* 187.

To relieve portal congestion; a useful purgative in general
plethora, dropsy, and in other conditions relieved by purgation,
particularly in gouty constitutions.

*Dose—*One every four hours until the bowels are well relieved.

(23.)
Colchici et Pil. Hydrarg.

℞ Ext. Colchici Acet., gr. j.; Ext. Aconiti Alc., gr. $\frac{1}{13}$;
Pil. Hydrarg., gr. iij. M. ft. pil.

For gout and rheumatism, with deficient action of the liver.
The efficacy of this pill is increased when followed by a dose of
Friedrichshalle water.

*Dose—*One or two at bedtime.

(24.)
Arsenici.

℞ Acid. Arseniosi, gr. v.; Pulv. Acaciæ, gr. xxx.; P.
Cinnam. Co., gr. xxx.; Ext. Jalapæ, gr. cxx. M. ft. mass,
et divid. in pilulæ 100.

In psoriasis and chronic eczema. Extensively used in India in the
treatment of lepra and other scaly diseases of the skin. May be
substituted in most cases for Fowler's Solution.

*Dose—*One three times a day.

In common use in the Skin Hospital.

See also No. 121 *and* 127.

(25.)
Argent. Nit. et Hyoscyami.

℞ Argent. Nit., gr. ½; Ext. Hyoscyami, gr. iij. M. ft. pil.

Nitrate of Silver has been found most useful in the treatment of
many obstinate forms of dyspepsia, by lessening the sensibility of
the nerves of the stomach. It has also been extensively employed
in diseases of the nervous system. Also in obstinate and chronic
forms of diarrhœa, and in the diarrhœa of typhoid.

*Dose—*One pill twice or three times a day. May be continued
for three or four weeks. If continued for a long period it is said to
discolour the skin. This never occurs in less than three months,
and it is not often desirable to give it more than half that time
without interruption.

See Nitrate of Silver with Opium, F. 37.

(26.)

Pot. Bromid. et Valerian.

℞ Potassii Bromid., gr. iv. ; Ext. Valerian., gr. j. M. ft. pil.

A useful form for prescribing the bromide when large doses are not indicated. A good remedy for hysteria and epileptic affections, especially in subjects exhibiting extraordinary excitement of the sexual organs. Bromide of ammonium may be used for the same purposes.

See F. 18 *and F.* 155.

Dose—Two three times a day, or oftener.

(27.)

Hyd. c. Cretâ et Rhei.

℞ Hyd. c. Cretâ, gr. ij. ; Ext. Rhei., gr. j. ; Ipecac., gr. ¼. M. ft. pil.

This and the following formula are well adapted for children. They are mildly aperient and alterative in their action, and they relieve the practitioner from prescribing powders which not only nauseate but excite little patients into rebellion. In order to reduce the bulk of the pill, Ext. Rhei, gr. j. (equal to at least three of the powders) has been substituted for the Pulv. Rhei. F. 28 is rather the more active of the two. (See Glycecols.)

Dose—One or two at bedtime.

See F. 188.

(28.)

Pil. Hydrarg. et Rhei.

℞ Pil. Hydrarg., Ext. Rhei, aa. gr. j. ; P. Ipecac., gr. ¼. M. ft. pil.

Dose—One or two at bedtime.

(29).

Pil. Hydrarg. c. Opio.

℞ Pil. Hydrarg., gr. iij. ; Pulv. Opii, gr. ½. M. ft. pil.

Useful in primary syphilis, and in other affections when it is desired to bring the system under the influence of mercury.

Dose—One, twice or three times a day, watching the gums.

(30.)
Pot. Iodid. et Colchici.

℞ Pot. Iodid., gr. iij. ; Ext. Colchici, B.P., gr. j. M. ft. pil.

A useful remedy in gout and rheumatism. The Mist. Alkalina Aromat. may be taken with advantage with these pills.

See F. 11.

(31.)
Zinci Sulph. et Belladonnæ.

℞ Zinci Sulph., gr. viij. ; Ext. Belladon., gr. ij. M. ft. gran. 8.

A very efficacious remedy for hooping-cough. Tasteless and small (mere granules). Children take them readily.

Dose—For a child above three years of age, one every six hours ; every other day the remedy may be increased by an additional dose, the action of the belladonna of course being watched. These granules are also valuable in incontinence of urine in childhood, irritability of urinary organs, etc.

See F. 136.

. (32.)
Quinæ Valer. et Quassiæ.*

℞ Quinæ Valer., gr. j. ; Ext. Quassiæ, gr. ij. M. ft. pil.

Very useful in hysteria and analogous nervous disorders; in facial neuralgia it is very efficacious.

Dose—One or two, three times a day, or oftener in severe cases.

(33.)
Stramonii et Belladonnæ.

℞ Ext. Stramonii, gr. ¼ ; Ext. Belladonnæ, gr. ¼. M. ft. gran.

In asthma, the combination of stramonium and belladonna often relieves when either drug administered separately fails.

Dose—One every four hours.

See also F. 152, 153, 158.

---

* In intermittent neuralgia, hemicrania, etc., Dr. Neligan says, "This is a very excellent preparation. It fulfils two effects very often indicated in this class of disease. It certainly is one of the most efficient of the chemical nutritive tonics which, in accordance with improved methods of treating chronic diseases, have become so desirable to the physician."

"Of great value in the treatment of all forms of strumous disease, and in general debility."—Dr. TANNER.

(34.)
Assafœtidæ et Valerian.

℞ Assafœtidæ Colat., gr. iij. ; Zinci Valer., gr. j. M. ft. pil.

A very useful medicine in hysteria.

Dose—Two or three twice or thrice a day.

(35.)
Acid. Gallic. et Cannabis Ind.

℞ Acidi Gallici, gr. iv.; Ext. Cannabis Ind., gr. j. M. ft. pil.

Astringent, anodyne. To check night-sweats in phthisis.

Dose—One at bedtime.

(36.)
Acid. Gallic. et Morphiæ.

℞ Acidi Gallici, gr. ijss.; Morphiæ Hydroch., gr. ₁₆. M. ft. pil.

Use as above, also in hæmoptysis and some other hæmorrhages.

Dose—Two or three every four or six hours.

Both these forms are used at the *Hospital for Consumption.*

See 130.

(37.)
Argent. Nit. c. Opio.

℞ Argent. Nit., gr. ½ ; Ext. Opii, gr. j. M. ft. pil.

A very useful remedy in obstinate chronic diarrhœa, chronic gastritis, etc. Checks pain and vomiting. *See Note to F.* 25.

Dose—One night and morning, or oftener.

(38.)
Cupri Sulph. c. Opio.
(*Guy's.*)

℞ Cupri Sulph., gr. ¼ ; Ext. Opii, gr. ½ ; Ext. Gentian., gr. j. M. ft. pil.

In obstinate diarrhœa, in phthisis, typhoid fever, etc.

Dose—One repeated every four hours.

(39.)

Hyd. c. Cretâ et P. Doveri.

℞ Hyd. c. Cretâ, gr. j. ; P. Ipecac. Co., gr. ½ ; Sodæ Carb. Ex., gr. j. M. ft. pil.

A useful remedy for infantile diarrhœa, etc. It promotes the action of the liver, and corrects acidity, relieves griping, and diminishes the frequency of the stools. Each pilule contains Opium gr. $\frac{1}{10}$.

Dose—According to age : one to a child six or eight months old.

See F. 92 for double strength.

(40.)

Assafœtidæ, Opii, et Capsici.

℞ Assafœtidæ Colat., gr. ij. ; P. Opii, gr. j. ; P. Capsici, gr. jss. M. ft. pil.

Useful in colic and affections of the bowels attended with pain and spasms.

See also 138.

(41.)

Hyd. Subchlor. et Scammon.

℞ Hyd. Subchlor., gr. j. ; P. Scammon. Virg., gr. ij. ; Jalapinæ, gr. jss. M. ft. pil.

A useful purge for children, especially if suspected to be suffering from intestinal worms.

Dose—One or two at bedtime.

See also following Formula and Glycecols.

(42.)

Hyd. Subchlor. et Scammon.

℞ Hyd. Subchlor., gr. j. ; P. Scammon. Virg., gr. iij.; P. Zingib., gr. j. M. ft. pil.

Use as above.

Dose—One or two at bedtime.

See also 51, *half strength.*

(43.)

Santonin et Scammon.

℞ Santonin, gr. ij.; Scammon. Res., gr. iv.   M. ft. pil. ij.

An excellent anthelmintic for children. Contains no mercury and may be repeated safely.

Dose—Two or three at bedtime on an empty stomach. A dose of castor oil should be given the following morning.

(44.)

Ferri et Quassiæ.

℞ Ferri Sulph. Exsic., gr. jss.; Ext. Quassiæ, gr. ij. M. ft. pil.

A very useful tonic for children suffering with ascarides. One pilule twice or thrice daily, with an occasional dose of F. 43.

It may be given to children of two years old and upwards.

(45.)

Pepsinæ et Aloes.

℞ Pepsinæ Porci, gr. iij.; Ext. Aloes Barb., gr. j. M. ft. pil.

An excellent remedy for atonic dyspepsia with constipation; also in certain forms of suppressed menstruation. Each pill is equal to twenty grains of the pepsine of commerce. Twelve dozen of these pills contain one ounce of pepsina porci; they are therefore *necessarily* expensive, though not so dear as formerly.

(46.)

Ipecac., Rhei, et Argent. Oxid.

℞ P. Ipecac. Ver., Argent. Oxid., aa. gr. j.; Ext. Rhei, gr. j. M. ft. pil.

A good dinner pill when there is a sense of oppression and uneasiness after food—the effect of slow digestion. The oxide of silver is said never to cause discoloration of the skin.

See Note to F. 25.

(47.)

Quinæ et Rhei.

℞ Quinæ Sulph., gr. j.; Pil. Rhei Co., gr. jss.; Ext. Lupuli, gr. viij. M. ft. pil. 2.

A useful tonic and mild aperient in many forms of dyspepsia.

Dose—One or two every day *with* dinner.

Pil. Pepsinæ Co. (48.)

Pil. Digestiv.

℞ Pepsinæ Porci, gr. j.; Ext. Rhei, Ext. Aloes Soc., P. Capsici, aa. gr. ½; P. Canellæ Cort., Ext. Gentian., aa. gr. j. M. ft. pil.

This pill has long been a favourite remedy for the most common forms of indigestion. The dose of pepsine has been increased which is calculated to materially add to its efficacy.

Dose—One or two with the principal meals.

(49.)
Bismuthi et Hyoscyami.

℞ Bismuth. Subnit., gr. ijss.; Ext. Rhei, Ext. Hyosc., aa. gr. j. M. ft. pil.

Useful in gastralgia, pyrosis, pleurodynia, etc.

Dose—Two or three *before* meals.

See Bismuth and Iron, F. 165.

(50.)
Sodæ, Rhei, et Chiratæ.

℞ Sodæ Carb., gr. ij.; Ext. Rhei, gr. ½; P. Zingib., gr. j.; Ext. Chiratæ, gr. j. M. ft. pil.

An excellent antacid, aperient, and tonic medicine for children as well as adults. A very useful remedy for stomach derangements occurring in children, accompanied with herpes, eczema.

Dose—One, two, or three, half an hour before food.

(51.)
Pil. Cathartic Co.

United States Pharmacopœia.

℞ Hyd. Subchlor., gr. j.; Ext. Coloc. Co., gr. i¼.; Ext. Jalapæ, gr. j.; P. Gambogæ, gr. ¼; Gingerinæ, gr. ₁′₂. M. ft. pil.

An excellent purgative, combining efficiency of action and comparative mildness with smallness of bulk. It is a capital antibilious pill. The following modified formula will commend itself to those who have experienced the inconvenience of administering bulky powders to children.

(51A.)

℞ Hydrarg. Subchlor., gr. ij. ; Ext. Aloes Pur., gr. ij. ; Ext. Jalapæ, gr. j. ; Gingerinæ, q.s. M. ft. gran. 2.

These pills, on account of their smallness, are known as Cathartic *Granules.* They act admirably on children of all ages. In cases where a mercurial is admissible, no better can be employed. Quite as efficient as calomel and jalap powder, and much more pleasant and convenient.

Dose—One or two, according to the age of the patient.

See also F. 66, *Aperient Granules.*

(52.)
Pil. Elaterii Co.

℞ Elaterii, gr. jss. ; Pulv. Capsici, gr. ix. ; Hyd. Subchlor., gr. xij. ; Ext. Hyoscy., gr. xviij. M. ft. pil. 12.

A good hydragogue cathartic. The capsicum prevents the nausea which elaterium so frequently excites.

Dose—Two or three.

(53.)
Podophylli et Rhei.

℞ Podophylli Res., gr. ¼ ; Pulv. Rhei, gr. ijss.; Ext. Hyoscy., gr. jss. ; P. Capsici, gr. ¼. M. ft. pil.

A useful alterative and mild aperient in jaundice from suppression, torpid liver, in dropsy from cardiac and renal or hepatic diseases.

Dose—Two every night at bedtime.

(54.)
Nucis Vomicæ et Rhei.

℞ Ext. Nucis Vom., gr. iij. ; Pulv. Ipecac., gr. vj. ; Pil. Rhei Co., gr. xl. M. ft. pil. xij.

In habitual constipation from atony of the coats of the bowels, with deficient secretions of intestinal mucus.

Dose—Two every other night at bedtime.

(55.)
Nucis Vomicæ et Coloc. Co.

℞ Ext. Nucis Vom., gr. ¼; P. Coloc. Co., gr. ij.; Ext. Hyoscy., gr. j.; Ext. Aloes Aquosi, gr. j. M. ft. pil.

(56.)
Pil. Crotonis Co.

℞ Ol. Crotonis, min. ij.; Pil. Coloc. Co., gr. xxx.; Pil. Assafœtidæ Co., gr. xxx. M. ft. pil. 12.

A brisk cathartic. Useful in cerebral congestion, apoplexy, and visceral obstructions. Also in cases of sciatica depending upon loaded colon, etc.

Dose—Two or three at bedtime for a few nights.

See F. 126.

(57.)
Fellis et Ammon. Carb.

℞ Fellis Bovis Pur., gr. xxxvj.; Ammon. Carb., gr. xxiv. M. ft. pil. 12.

Useful in some forms of functional dyspepsia, especially where vomiting occurs after food.—DR. TANNER.

(58.)
Fellis et Coloc.

℞ Fellis Bovis, aa. gr. xx., Pil. Coloc. et Hyosc., Ext. Lupuli. M. ft. pil. 12.

A mild laxative and tonic; may be employed in the same case as that recommended in formula 57, attended with deficient excretion of biliary matter.

(59.)
Coloc. et Assafœtidæ.

℞ Pil. Coloc. et Hyoscy., Pil. Assafœtidæ Co., aa. ijss. M. ft. pil.

A useful aperient in hysteria with flatulence.

(60.)
Pil. Coloc. Hyosc. et Podoph.

℞ Pil. Coloc. et Hyoscy., gr. iv. ; Podophylli Res., gr. ¼.
M. ft. pil.

A very favourite formula. A certain and safe cathartic without
mercury. A good pill to entrust to patients for occasional use as
aperient or antibilious medicine.

(61.)
Pil. Hydrarg. et Coloc.

℞ Pil. Hydrarg., Pil. Coloc. Co., aa. gr. ij. M. ft. pil.

A mild aperient and antibilious pill.

See Blue Pill, Colocynth, and Henbane, F. 141.

(62.)
Cal. c. Coloc.

℞ Hyd. Subchlor., gr. j. ; Pil. Coloc. et Hyoscy., gr. iv.
M. ft. pil.

A good form of giving calomel as a cholagogue ; purgative. The
Henbane prevents griping.

Dose—Two at bedtime.

(63.)
Hydrarg. Coloc. et Rhei.

℞ Pil. Hydrarg., gr. ½ ; Pil. Coloc. Co., Pil. Rhei Co., aa.
gr. j. M. ft. pil.

A very useful *little* pill. Acts exceedingly well upon children and
aged persons.

Dose—One, two, or three, at bedtime.

For the same pill, double strength, see F. 142.

(64.)
Quinæ et Nucis Vom.

℞ Quinæ Sulph., gr. j. ; Pil. Rhei. Co., gr. ij. ; Pulv. Capsici,
gr. ½ ; Ext. Nucis Vom., gr. ¼. M. ft. pil.

An excellent tonic and mild laxative. Useful in a host of cases
in which quinine is indicated.

(65.)
Coloc. Co. et Rhei. Co.

℞ Pil. Coloc. Co., Pil. Rhei Co., aa. gr. ijss. M. ft. pil.

A favourite combination of some practitioners.

(66.)
Jalapinæ et Aloes.

℞ Ext. Aloes Soc., gr. ½; Jalapinæ, gr. ½; Pulv. Ipecac. gr. ¼; Saponis Mollis, q.s. M. ft. pil.

An effective aperient for infants and children.

See also 51A *and Aloetic Granules, F.* 171.

(67.)
Aloes et Nucis Vomicæ.

℞ Ext. Aloes Soc., gr. j.; P. Ipecac., gr. ¼; Ext. Nucis Vom., gr. ₁₂. M. ft. pil.

Jalapine, formerly in this formula, is now omitted, and the pure Extract of Aloes used,—thus the efficiency of this pilule has been increased, and its bulk reduced considerably. A very suitable medicine for children suffering with habitual constipation.

(68.)
Aloes et Myrrh. c. Ferro.

℞ Pil. Aloes et Myrrhæ, gr. iij.; Ferri Sulph. Exsic., gr. jss; Ext. Nucis Vom., gr. ¼. M. ft. pil.

This form has been improved by the addition of a small dose of Nux. It will be found exceedingly useful in chlorosis, amenorrhœa, hysteria, debility, etc.

Dose—One twice or thrice daily after food, or two every night at bedtime.

(69.)
Ferri Iodid. et Assaf.

℞ Pil. Ferri Iodid., gr. iij.; Pil. Aloes et Assafœtidæ, gr. iij.; Ext. Aloes Barb., gr. j. M. ft. pil. 2.

A useful tonic, alterative and laxative, in the treatment of chlorosis and amenorrhœa in strumous patients.

Dose—One or two, three times a day, with food.

C

(70.)

Quinæ, Ferri, et Hyoscyami.

℞ Quinæ Sulph., Ferri Sulph. Ex., Ext. Hyoscy., aa. gr. xx.
M. ft. pil. 12.

A favourite combination,—useful in debility and irritability of
the nervous system.

Dose—One twice a day.

(71.)

Quinæ et Belladonnæ.

℞ Quinæ Sulph., gr. xxiv.; Ext. Belladonnæ, gr. iv.; Camphoræ, gr. xxx. M. ft. pil. 12.*. DR. TANNER.

A capital tonic and sedative in painful affections, neuralgia, dysmenorrhœa, cancer, etc., when a sedative and tonic are needed.

Dose—One pill twice or three times a day.

See also No. 21.¶

(72.)

Zinci et Quassiæ.

℞ Zinci Sulph., gr. j.; Ext. Quassiæ, gr. ij. M. ft. pii.

A very useful tonic in cases where iron is not well borne.

Dose—One twice or three times a day.

(73.)

Quinæ et Ipecac.

℞ Quinæ Sulph., gr. j.; Pulv. Ipecac., gr. j.; Ext. Gentian.,
gr. ij. M. ft. pil.

See F. 20.

In cases of slow digestion.

Dose—One with meals.—DR. TANNER.

---

* It is desirable to divide these quantities into 24 instead of 12 pill . Fo
general purposes the dose given above is too large.

(74.)
Ferri Carb. c. Quinâ.

℞ Quinæ Sulphatis, gr. j.; Pil. Ferri Carb., gr. iij. M. ft. pil.
Useful in anæmia and chlorosis.
Dose—One twice or thrice daily.

(75.)
Quinæ et Nucis Vomicæ.

℞ Quinæ Sulph., gr. xviij. ; Ext. Nucis Vom., gr. iij. ; Ext.
Gentian., gr. xviij.; Pulv. Capsici, gr. vj. M. ft. pil. 12.

A very favourite combination, and good form of exhibiting
quinine in debility and constipation.—DR. TANNER.

(76.)
Pepsinæ et Ferri.

℞ Ferri Redacti, gr. xxxvj. ; Pepsinæ Porci, gr. xxxvj. ;
Zinci Phosphatis, gr. xviij. M. ft. pil. 24.

Recommended by Dr. Tanner in anæmia, etc., with weakness of
the digestive organs.

See Note to F. 45.

(77.)
Strychniæ et Ferri.

℞ Ferri Redacti, gr. xl. ; Zinci Valer., gr. xx. ; Strychniæ,
gr. j. M. ft. pil. 20.

In hypochondriasis and great nervous depression.

See F. 85; also Pil. Phosphori Comp., F. 200.

(78.)
Zinci Valer. et Quinæ.

℞ Zinci Valer., gr. xij. ; Quinæ Sulph., gr. xij. ; Pil. Rhei
Co., gr. xviij.; Ext. Anthemidis, q.s. M. ft. pil.

A useful combination in debility, with hysteria, neuralgia, etc.
Dose—One three times a day.

(79.)
Zinci Sulph. et Aconiti.

℞ Zinci Sulph., gr. ij.; Ext. Aconiti, B.P., gr. j.; Ext. Quassiæ, q.s. M. ft. pil.

A good nervine, tonic, and astringent, useful in epilepsy, neuralgic pains, lumbago, pleurodynia, etc.

Dose—One three times a day.

(80.)
Anti Malarial Pill.

℞ Quinæ Sulph., gr. j.; Ferri Sulph. Exsic., gr. ¼; Ol. Res. Pip. Nig., gr. ₁/₁₆; Acidi Arseniosi, gr. ₁/₈₀; Podophyllin, gr. ⅛. M. ft. pil.

Dose—Two three times a day.

See Dr. Livingstone's *Pill, F.* 178.

(81.) *
Ferri, Quinæ, et Nucis Vom.
(*Pil. Tonicæ c. Quinâ.*)

℞ Quinæ Sulph., gr. xij.; Ferri Sulph. Ex., gr. xij.; Ext. Nucis Vom. gr. vj.; Ext. Quassiæ, gr. vj. M. ft. pil. 12.

A very favourite formula and largely prescribed. A happy combination, which fills many indications.

Dose—One or two twice or three times daily with food.

(82.)
Ferri Iodid. et Quinæ.

℞ Pil. Ferri Iodid., gr. jss.; Quinæ Sulph., gr. ½. M. ft. pil.

A useful tonic in debility for strumous children. Also in amenorrhœa and chlorosis; in chronic rheumatism; in goître and other glandular swellings.

(83.)
Zinci Sulph. et Calumbæ.

℞ Zinci Sulph., gr. j.; Ext. Calumbæ, gr. ij. M. ft. pil.

A useful tonic in some stomach derangements.

Dose—One twice or three times a day before food.

* For a similar but less expensive combination, see F. 173.

(84.)

Aconiti et Quinæ.

℞ Ext. Aconiti Alcohol, gr. ₁'₆.; Quinæ Sulph., gr. j. M.
ft. gran. DR. PROSSER JAMES.

These granules are exceedingly valuable in tic and other *acute*
neuralgic affections, giving speedy relief to pain and soothing the
general system.

Dose—One may be taken and repeated in an hour, after which one
every two, three, or four hours. In a severe case two may be taken
for the first dose. As soon as the effects of the aconite are observed, they should be omitted and pills of quinine alone substituted.

(85.)

Ferri Hypophos. et Strychniæ.

℞ Ferri Hypophosphitis, gr. xl. ; Strychniæ, gr. j. ; Ext.
Quassiæ, q.s. M. ft. pil. 20.

A very good "pick up." In hypochondriasis, neuralgia, and
mental depression a favourite tonic. *It is a very small pill,* and
frequently prescribed for ladies who object to bulky medicines.

(86.)

Ferri, Quassiæ, et Quinæ.

℞ Ferri Sulph. Exsic., gr. j. ; Ext. Quassiæ, gr. j. ; Quinæ
Sulph., gr. ¼. M. ft. pil.

A very useful tonic for delicate children, especially those suffering
with ascarides.

(87.)

Ferri, Galbani, et Coloc.

℞ Ferri Sulph. Ex., Pil. Galban. Co., aa. gr. jss.; Pil.
Coloc. Co., gr. j¼. M. ft. pil.

A capital pill for hysterical and nervous females suffering with
irregular menstruation, costive bowels, flatulence, etc.

(88.)

Hyosc., Camphor. et Lupuli.

℞ Ext. Hyoscyami, Camphoræ, Lupulinæ, aa. gr. xx. M. ft. pil. 12.

A very useful general sedative and hypnotic. Prescribed with advantage for hysterical and hypochondriacal patients suffering with sleeplessness. Also in cases where opium and its compounds are not well borne.

Dose—Two at bedtime.

(89.)

Camphoræ et Belladonnæ.

℞ Camphoræ, gr. v. ; Ext. Belladonnæ, gr. ½ ; Ext. Hyosc., gr. iij. M. ft. pil. 2.

A useful antispasmodic and sedative in spermatorrhœa, chordee, and to relieve spasms of the air passages, etc., irritability of urinary organs, and in spasmodic cough.

Dose—Two at bedtime.

(90.)

Aconiti et Opii.

℞ Ext. Opii Pur., gr. j. ; Ext. Aconiti, gr. j. M. ft. pil.

(*Note.*—The formula in former editions, numbered 90, had no therapeutic value above that of Dover's powder, and has been omitted.)

Useful in acute inflammations, peritonitis, pleurisy, ovaritis, etc.

Dose—One every four, six, or eight hours.—DR. TANNER.

(91.)

Opiate Granules for Infants.

℞ Pulv. Ipecac. Co., gr. j. ; Sacch. Lactis, gr. iij. M. ft. gran. 4.

"May be given safely to infants from two to six weeks old."—DR. TANNER.

Dose—One dissolved in warm milk at night, or when required.

(92.)
Hyd. c. Cretâ et P. Doveri.

℞ Hyd. c. Cretâ, gr. ij. ; P. Ipecac. Co.,'gr. j.   M. ft. pil.

Alterative, sedative, and diaphoretic. A useful combination in the treatment of inflammatory affections of the mucous surfaces, especially of the bowels—enteritis, dysentery, and some forms of diarrhœa.

Dose—One every few hours, followed by a small dose of castor oil.

See also Nos. 39 *and* 149.

(93.)
Pot. Bromid. et Belladon.

℞ Pot. Bromid., gr. ij.; Ext. Belladonnæ, P. Ipecac., aa. gr. ¼. M. ft. pil.

A useful remedy for hooping-cough.

See also F. 31 ; *Glycecols.*

(94.)
Codeiæ et Assafœtidæ.

℞ Codeiæ, gr. ½ ; Pil. Assafœtidæ Co., gr. iv.   M. ft. pil.

Especially useful in spasmodic cough and dyspepsia.

(95.)
Opii et Belladonnæ.

℞ Pulv. Opii, gr. j. ; Ext. Belladonnæ, gr. ½.   M. ft. pil.

Valuable in cases where it is desirable to relieve pain by opium without inducing constipation.

Dose—One every six hours.

The Belladonna overcomes the astringency of the Opium without destroying its anodyne effects.

Opium and Belladonna are said to be *antagonistic* in their effects. Notwithstanding, this is a very favourite formula, and prescribed with great advantage.

(96.)
Morphiæ, Scillæ, et Ipecac.

(Cough Granules.)

℞ Morphiæ Hydroch., gr. $\frac{1}{30}$; Pulv. Ipecacuanliæ, gr. $\frac{1}{12}$;
Bals. Tolu, gr. $\frac{1}{4}$; Pulv. Scillæ, gr. $\frac{1}{12}$; Sacch. Alb., q.s.
M. ft. pil.

These granules are intended as substitutes for Trochisci Morphiæ
et Ipecac. of the P.B. The addition of the Squills it is thought
will add efficiency to their expectorant properties, and render
them available for a larger number of cases. Each granule
contains gr. $\frac{1}{30}$ Morphia, and gr. $\frac{1}{12}$ each of Squills and Ipecacuan.
One, therefore, would be a dose for a child 2 years old, two for
5 to 7 years, three for 9 years, while the adult dose would be from
two to six, according to the requirements of the case and the frequency of the repetitions. They will be found exceedingly useful
in the treatment of almost every kind of cough.

See also Pil. Scillæ c. Morphiâ, F. 117.

(97.)
Camphoræ Comp.

(Paregoric Granules.)

℞ Pulv. Opii, gr. xviij. ; Acid. Benzoici, gr. xviij. ; Camphoræ, gr. xv. ; Ol. Anisi, gr. xij. ; Pulv. Altheæ, gr. clxxx.
M. ft. gran. 288.

Each granule contains gr. $\frac{1}{16}$ of P. Opii, and is equal to min. xv.
of *Tinct. Camph. c. Opio.*

(98.)
Pil. Salinæ et Doveri.

℞ Antim. Pot. Tart., gr. $\frac{1}{8}$; P. Ipecac. c. Opio, gr. ijss. ;
Pot. Nitratis Exsic., gr. ij. M. ft. pil.

The following combinations of antimony, Dover's powder, and
nitre have been devised with a view to supply useful diaphoretic
sedatives in a *portable* form, so that the practitioner having them
always with him on his rounds, would be enabled to administer
suitable remedies in such acute cases as he may be called upon to

prescribe for. Useful in *inflammation of the mucous membranes, catarrh, tonsillitis, bronchitis;* also in *acute rheumatism* and various *febrile states.*

The Dover's powder in all these preparations is made with Potassæ Nitras in place of Potassæ Sulphas. "They are preferable to the officinal powders, as the nitrate acts better than the sulphate." —Dr. TANNER.

Dose—One or two pills every three, four, or six hours.

> *See Mist. Diaphoretica and Mist. Ammon. Acet.*
> *Also Pil.* „ *F.* 165.

(99.)
Pil. Salinæ et Ipecac.

℞ Antim. Pot. Tart., gr. $\frac{1}{16}$; P. Ipecac., gr. $\frac{1}{2}$; P. Pot. Nitratis Ex., gr. iv. M. ft. pil.

A useful expectorant and diaphoretic in *catarrh, bronchitis,* etc.

Dose—One or two every two, three, or four hours, according to age, sex and the effect desired.

(100.)
Antim. P. Tart. et P. Doveri.

℞ Antim. Pot. Tart., gr. $\frac{1}{4}$; P. Ipecac. c. Opio, gr. v. M. ft. pil.

Double the strength of F. 98.

A very useful sedative diaphoretic.

Dose—One every three, four, or six hours. Will be well borne after a few doses of the milder preparation.

> *See F.* 98.

(101.)
Digitalis et Scillæ.

(*Consumption*).

℞ Pulv. Digitalis, Pulv. Scillæ, aa. gr. j.; Ext. Conii., gr. ij. M. ft. pil.

Sedative and diuretic. In dropsy, heart disease, etc.

Dose—One or two twice or thrice daily.

(102.)

Ipecac., Pot. Nit., et Papav.

℞ Pulv. Ipecac., gr. ½ ; Pot. Nitratis, gr. iij. ; Ext. Papav., gr. ¼. M. ft. pil.

A simple expectorant, diaphoretic, mildly depressant, and ano-dyne. Very useful for children, in catarrh and bronchitis, pneu-monia, and in febrile conditions depending on painful dentition. A good substitute for saline expectorant mixtures.

Dose—For an infant *one*, for a child two to five years old *two*, repeated every two, three, four, or six hours, according to age and symptoms, dissolved in warm, milk or gruel if it cannot be given whole. *See Glycecols.*

(103.)

Pot. Chlór. c. Ipecac. et Aconiti.

℞ Pot. Chlor., gr. iv. ; Ipecac., gr. ½ ; Ext. Aconiti Alco-holic, gr. ₁'₂. M. ft. pil.

Chlorate of Potash in its therapeutic action resembles that of Nitre in being refrigerant and diuretic; but it has also a special action of its own on the mucous membrane and is not depressant. Indeed, it is believed by many to be a restorative, acting by giving up 'ts large supply of oxygen to the blood. It is, therefore, to be pre-'erred in low fevers, scarlatina, typhus, etc. The Aconite gives to this remedy sedative and diaphoretic properties. It will be found highly useful in low forms of inflammation, sore throat, sub-acute rheumatism, etc.

Dose—One pill frequently repeated every three or four hours until the effect of the aconite is produced.

(104.)

Digitalis et Hyd. Subchlor.

(*St. Thomas's Hospital.*)

℞ P. Digitalis, Hyd. Subchlor., P. Scillæ, aa. gr. j. ; Ext. Hyosc., gr. jss. M. ft. pil.

A useful combination, and, when Calomel is not contra-indicated, the best form to prescribe, as the Calomel *increases* the diuretic power of digitalis and squill.

See also No. 101 and 9.

(105.)

Ferri Fœtidæ.

(St. Bartholomew's.)

℞ Ferri Carb. Sacch., gr. iij. ; Pil. Assafœtidæ Co., gr. ij. M. ft. pil.

A useful remedy in hysteria, well adapted for young persons who do not bear the stronger preparations of iron.

Dose—One or two three times a day.

(106.)

Aloes c. Ferro et Nucis Vom.

℞ Pil. Aloes c. Ferro, gr. iv. ; Ext. Nucis Vom., gr. ¼. M. ft. pil.

The addition of Nux Vomica to the Pharmacopœial preparation is considered to increase its emmenagogue properties.

(107.)

Emmenagog. Comp.

℞ Ferri Sulph. Exsic., gr. xxiv. ; Aloes Soc., gr. vj. ; P. Canellæ, gr. xij. ; Ol. Sabinæ, min. vj. ; Terebinth., U.S.P.; q.s. M. ft. pil. xviij.

A useful stimulating emmenagogue. Employed with advantage in suppressed, absent, or deficient menstruation, after a general plan of treatment has been adopted for the removal of any morbid state of the system,—anæmia, plethora, etc.

(108.)

Pil. Copaibæ Comp.

℞ Ol. Copaibæ, mj. ; P. Cubebæ, gr. ij. ; Terebinth. Alb., q.s.; Ferri Sulph., gr. ½. M. ft. pil.

In gonorrhœa and gleet.

Dose—Two or three twice or thrice daily.

(109.)
Sodæ Carbonatis.

R Sodæ Carb. Exsic., gr. xlij.; Ol. Carui, min. iij.; Pulv. Zingib., gr. vj. ; Saponis, gr. vj. M. ft. pil. 12.

Antacid and corrective in heartburn, flatulence, etc. A useful form for prescribing an alkali when a solution would be inconvenient.

See Bicarbonate of Potash with Colchicum, F. 159 and 50.

(110.)
Hydrarg. Subchlor. et P. Doveri.

R Hyd. Subchlor, gr. j.; P. Ipecac. c. Opio, gr. iv. M. ft. pil.

A very useful combination of calomel and opium. Acting as a sedative, alterative, and diaphoretic. In this form it is to be preferred to the formulæ 1 to 6, when it is desired to obtain the alterative rather than the specific action of the calomel.

Useful in inflammations of the mucous tract and other febrile conditions.

Dose—One every four or six hours, watching the gums; in mild cases, one or two at bedtime.

(111.)
Aconiti et Ipecac.

R Ext. Aconiti Alc., gr. jss.; Pulv. Ipecac., gr. vj.; Ext. Glycyr., q.s. M. ft. pil. 12.

In the treatment of phthisis when an expectorant is needed and the pulse is high, the combination of Ipecac. with Aconite will be found of service. This is also a good pill in acute bronchitis.

(112.)
Hyd. Subchlor. et P. Antim.

R Pulv. Antimonialis, gr. ij.; Hyd. Subchlor., gr. j. M. ft. pil.

Sudorific and resolvent. In inflammatory affections.

Dose—One for children, two for adults.

(113.)
Ferri et Chiratæ.

℞ Ext. Chiratæ, gr. ij.; Ferri Sulph. Ex., gr. j. M. ft. pil.

A very useful tonic.

Dose—One or two with meals.

(114.)
Colchici et P. Doveri.

℞ Ext. Colchici Acet., gr. j.; Pulv. Ipecac. c. Opio., gr. iij. M. ft. pil.

Antarthritic and sudorific. A valuable combination; found efficacious in both acute and chronic cases. Similar formulæ used at *King's College, St. George's, Middlesex*, and *London, Hospitals*. Known as Dr. Vance's Gout and Rheumatic Pills.

(115.)
Colchici et Belladon.

℞ Ext. Colchici Acet., gr. j.; Ext. Belladonnæ, gr. ½; Ext. Anthemidis, q.s. M. ft. pil.

Antarthritic and anodyne. *St Thomas's Hospital.*

(116.)
Colchici et Colocynth.

℞ Ext. Colchici Acet., gr. j.; Pil. Coloc. Co., gr. ij.; Ext. Belladonnæ, gr. ¼. M. ft. pil.

A useful purgative in gout and rheumatism.

Dose—One twice or thrice daily, according to the action on the bowels.

(117.)
Scillæ et Morphiæ.

℞ Pil. Scillæ Co., gr. ijss.; Morphiæ Hydroch., gr. $\frac{1}{12}$. M. ft. pil.

Expectorant anodyne. A very useful cough pill; small and very efficient.

Dose—One twice a day, and *two* at bedtime.

(118.)
Aloes et Nucis Vom.

℞ Ext. Aloes Soc., gr. ij.; Ext. Nucis Vom., gr. ¼; Ext. Hyoscyami, gr. j.; Saponis Dur., gr. j. M. ft. pil.

A very useful pill in some forms of constipation.

See note to F. 169.

Dose—One at bed-time or after dinner.

A similar formula is used at *St. Thomas's* and *University College* hospitals.

(119.)
Podophylli et Rhei.

℞ Podophylli Res., gr. ¼; Pil. Rhei Co., gr. iij.; Ext. Hyosc., gr. j. M. ft. pil.

Successfully prescribed in constipation depending on a diminished excretion of bile.

One pill taken every morning or every other morning, with breakfast, usually acts mildly and efficiently the *following morning*.

Dose—One or two.

See also Nos. 10, 53.

(120.)
Quassiæ, Zinci, et Galbani.

℞ Ext. Quassiæ, gr. jss.; Zinci Sulph., gr. j.; Pil. Galban. Co., gr. ij. M. ft. pil.

A useful tonic and antispasmodic in cases where the salts of iron disturb digestion.

(*See Galbanum and Quassia, F.* 87.)

(121.)
Pil. Asiaticæ.

℞ Acid Arseniosi, gr. lx.; Pip. Nig., ℥ix.; Acaciæ, q.s. M. ft. pil. 800.

"An excellent combination, highly esteemed in the East Indies as a remedy for lepra, psoriasis, and syphilitic eruptions; especially useful in languid habits of body."—NELIGAN. 702.

Dose—One or two daily.

See also Nos. 24 *and* 127.

(122.)

Antineuralgic Pills.

(*Dr. Gross, Philadelphia.*)

℞ Quinæ Sulph., gr. ij.; Morphiæ Sulph., gr. ₃¹₀ ; Ext. Aconiti Alc., gr. ₂¹₀ ; Strychniæ, gr. ₃¹₀ ; Arsenici Alb., gr. ₂¹₀. M.ft.pil.

Dose—One every four or six hours. A powerful remedy; given with caution, it is safe and highly successful.

(123.)

Opii, Camphoræ, et Ipecac.

The same form as 97, with the addition of Ipecac., gr. ½.

A very useful anti-cough medicine, possessing all the virtues of Tinct. Camph. c. Opio, plus those of Ipecacuan.

Dose—One every three, four, or six hours.

(124.)

Hyd. Iodid. Vir. et Sodæ.

℞ Hyd. Iodid. Vir., gr. j. ; Sodæ Carb. Exsic., gr. xvj. ; Pulv. Cretæ Aromat., gr. xvj. ; Ext. Sarsæ, q.s. M. ft. pil. 16.

As an alterative in the treatment of skin diseases with syphilitic taint.

Dose—One or two twice or three times a day, according to age.

(125.)

Quinæ et Ammoniæ.

℞ Quinæ Sulphatis, gr. j. ; Ammoniæ Carb., gr. iij. M. ft. pil.

In cases of debility, low fever, and of great exhaustion, in all cases where bark and ammonia are useful.

(126.)

Pil. Crotonis Co.

(*London Ophthalmic* and *Middlesex Hospitals.*)

℞ Ol. Crotonis, m iij. ; Ext. Coloc. Co., gr. lx. M. ft. pil. · xij.

A powerful cathartic, acting quickly.

Dose—Two, may be repeated if needful.

See F. 171.

(127.)

Pil. Arsenici Rub.

(*Dr. Wickham's Formula.*)

℞ Arsenici Alb., gr. vj.; P. Glycyrrhizæ, gr. xxx.; Antim. Sulph., gr. xc.; Ext. Gent., gr. lx.; Saponis Dur., gr. xx. M. secundum artem, et ft. pil. 48.

Each pill contains ⅛ of a grain of arsenic.

Dr. Wickham has employed this formula for forty years in scaly affections of the skin, with the utmost success. Although the dose, gr. ⅛, is apparently a large one, he has never seen any bad result from it.

Dose—One three times a day with meals.

This pill should be coated in a transparent coating, and its colour serves to distinguish it from all others, so that a mistake cannot occur.

(128.)

Podophylli et Quinæ.

℞ Podophylli Res., gr. ij.; Quinæ Sulphatis, gr. xij.; Pulv. Capsici, gr. vj.; Ext. Hyoscyam., gr. xij. M. ft. pil. xij.

This is an excellent form for combining the alterative effects of Podophyllin with the tonic properties of Quinine.

A very useful remedy for jaundice or suppression of bile in patients from tropical climates.

Dose—One or two three times a day.

(129.)

Podophyllin et Belladon.

(*American.*)

℞ Podophylli Res., gr. ½; Ext. Belladonnæ, gr. ⅛; Pulv. Capsici, gr. j.; Sacch. Lactis, gr. j. M. ft. pil.

Very useful in some forms of habitual constipation.

(130.)

Acid. Gallic. c. Opio.

℞ Acidi Gallici, gr. iv.; P. Opii, gr. ¼. M. ft. pil.

A useful astringent in phthisis; to check night sweats, hæmoptysis, and other hæmorrhages.

Dose—Two for first dose, and repeat one frequently.

(131.)

Aloes et Nucis Vom. c. Hyosc.

(*St. Thomas's Hospital.*)

℞ Ext. Aloes Soc., gr. j. ; Ext. Nucis Vom., gr. ½ ; Ext. Hyoscyami, gr. j. ; Saponis Dur., gr. j. M: ft. pil.

This and the following are very useful remedies for habitual constipation.

See also Nos. 132, 118, 168; *and Note to F.* 169.

(132.)

Aloes, Ferri, et Quinæ.

℞ Ext. Aloes Pur., gr. j. ; Ferri Sulph. Exsic., gr. j.; Quinæ Sulph., gr. ½. M. ft. pil.

See Note, F. 168 *and* 169.

(133.)

Hyd. Subchlor. et Antim. Tart.

℞ Hyd. Subchlor., gr. j.; Antim. Pot. Tart., gr. ¼. M. ft. pil.

This and the two following formulæ are active agents, invaluable in the treatment of sthenic inflammations, especially of the serous membranes. They should find a place in the "Pocket Vade Mecum" of every practitioner. They occupy very little space, and when required are powerful to relieve suffering and to arrest disease.

(134.)

Hyd. Subchlor., Antim. Tart. c. Opio.

℞ Hyd. Subchlor., gr. j. ; Antim. Tart., gr. ½ ; P. Opii, gr. ½. M. ft. pil.

(135.)

Antim. Tart. c. Opio.

(*Guy's.*)

℞ Antim. Tart., gr. ½ ; Pulv. Opii, gr. ½. M. ft. granul.

D

(136.)

Belladon. et Ipecac.

℞ Ext. Belladonnæ, gr. ¼; Pulv. Ipecac. Ver., gr. ½. M. ft. granul.

A useful remedy for hooping-cough, and may be given to young children alternately with Formula 31.

(137.)

Calomel et Coloc. Aperient Granules.

℞ Hyd. Subchlor., gr. ¼; Ext. Coloc. Co., gr. ¾. Ft. granul.

For this formula we are indebted to Dr. Ballard, who has found it a most efficient aperient for young children. They are exceedingly small, and are easily administered even to infants.

Dose—One for an infant, two for young children.

For a more active medicine see F. 51 A.

(138.)

Camphor. Capsici, c. Opio.

℞ Camphor, gr. ij.; P. Capsici, gr. jss.; P. Opii, gr. j. M. ft. pil.

A useful "ready remedy" for cholera and diarrhœa, and generally to relieve pain and spasm. It has been extensively used, both at home and abroad, and found to be exceedingly successful. Its portability is a great merit. As a stimulating anodyne astringent this remedy should be found in every medicine chest.

Dose—In severe cases two for the first dose, and one every two, three, or four hours afterwards. Two or three doses generally suffice. When not contra-indicated, a *wineglassful* of hot brandy and water is desirable after each dose.

See also No. 40.

(139.)

Cannabis Ind. c. Opio.

℞ Ext. Cannabis Ind. Alc., Pulv. Opii, aa., gr. ½. M. ft. gran.

Dose—One.

A very useful combination, to relieve pain and to procure sleep.

(140.)
Cannabis Ind., Belladon., et Ipecac.

℞ Ext. Cannabis Ind. Alc., gr. ¼ ; Ext. Belladonnæ, gr. ½ ; Ext. Opii, gr. ⅛ ; Pulv. Ipecac., gr. ¼. M. ft. gran.

Dose—One twice a day.

(141.)
Coloc., Hydrarg., et Hyosc.

℞ Pil. Coloc. Comp., gr. ij. ; Pil. Hydrarg., Ext. Hyoscyam., aa. gr. jss. Ft. pil.

A very favourite combination ; operates mildly, without griping.

(142.)
Coloc., Hydrarg., et Rhei.

℞ Pil. Coloc. Co., Pil. Hydrarg., Pil. Rhei Co., aa. gr. x. M. ft. pil. 6.

Equal parts of Compound Colocynth and Compound Rhubarb (F. 65) is with many practitioners a very favourite aperient pill. The addition of Blue Pill gives the formula a larger scope of usefulness. In this form it is a capital antibilious pill. For delicate and young persons five grains in two pills is a very efficient medicine.

See Formulæ 63 *and* 147.

(143.)
Ferri Hypophos. et Quinæ.

℞ Ferri Hypophosphitis, gr. iij. ; Quinæ Sulph., gr. j. ; Ext. Nucis Vom., gr. ⅛. M. ft. pil.

Dose—One, three times a day.

This is one of the best forms in which hypophosphite of iron and quinine can be administered.

(144.)
Ferri et Quinæ Sulph.
(*Chest*).

℞ Quinæ Sulph., Ferri Sulph., aa. gr. j. ; Ext. Anthem., gr. j. M. ft. pil.

See also Nos. 81, 70 *and* 173.

Dose—One or two, three times a day.

The two sulphates act well together in numerous cases.

(145.)

Ferri Valer. et Quinæ.

℞ Ferri Valer., gr. j. ; Quinæ Sulph., gr. ½. M. ft. pil.

Valerianate of iron is most useful in hysterical anæmic patients.

Dose—One, twice or three times a day.

(146.)

Hyd. Perchlor., Belladon., et Quinæ.

℞ Hydrarg. Perchlor., gr. ₁⁄₁₆ ; Ext. Belladonnæ, gr. ⅛; Quinæ Sulph., gr. ½. M. ft. pil.

In confirmed constitutional syphilis, as well as in some forms of eczema and other skin affections. A powerful alterative and tonic in disorders dependent on venereal taint.

Dose—One, three times a day, gradually increased to six pills daily.

In combination with Bark, Perchloride of Mercury was prescribed by Sir W. Wild ; and although said to be incompatible, it has been found to act very satisfactorily. The substitution of Quinine, form-ing a compatible preparation, is an obvious advantage.

(147.)

Hydrarg. et Rhei. Co.

℞ Pil. Hydrarg., Pil. Rhei Co., aa. gr. ij.; Ext. Hyoscyami gr. j. M. ft. pil.

See also Nos. 63 and 142.

A favourite combination.

(148.)

Hydrarg. et P. Doveri.

℞ Pil. Hydrarg., gr. ij.; P. Ipecac. c. Opio, gr. iij. M. ft. pil.

See also Nos. 39, 92, and 149.

Milder than combinations of Calomel and Dover's powder.

(149.)

Hyd. c. Cretâ et P. Doveri.

℞ Hyd. c. Cretâ, P. Ipecac. c. Opio, aa. gr. ijss. M. ft. pil.

See also Nos. 39, 92, and 148.

Still milder than the preceding form.

(150.)

Hyd. c. Cretâ et Quinæ.

℞ Hyd. c. Cretâ, gr. ij.; Quinæ Sulph., gr. ½; P. Rhei, gr. ij. M. ft. pil.

A very useful combination of bittter and nauseous medicines in a pleasant form, especially well adapted for children. In strumous ophthalmia and other forms of scrofula, where there is defective action of liver and other secretions, it will be found a good corrective.

Dose—One, three times a day.

For a similar remedy without grey powder, see No. 157.

(151.)

Aloes, Mastic, et Ipecac.

Pil. Prandii.

℞ P. Aloes Soc., gr. jss.; P. Mastic, gr. jss.; P. Ipecac., gr. j.; Ol. Carui, m ¼. M. ft. piL

A good anti-dyspeptic pilL

(152.)

Stramonii et Hyoscy.

℞ Ext. Stramonii, gr. iij.; Ext. Hyoscyami, gr. xx.; Ext. Lupuli, gr. xl. M. ft. pil. 12.

"In chronic disorders attended with suffering, in diseases of the nervous system, with pain and restlessness, and in the dyspnœa of phthisis and emphysema."—DR. TANNER.

Dose—One every four hours until relief is obtained.

(153.)

Morphiæ et Stramonii.

℞ Morphiæ Hydroch. gr. ⅛; Ext. Stramonii, gr. ¼; Pulv. Lupuli, gr. ij. M. ft. pil.

Use, the same as above. And is a more powerful sedative, from the presence of the morphia.

(154.)

Nucis Vomicæ, Coloc., et Rhei.

(London Hospital.)

℞ Ext. Nucis Vom., gr. ¼ ; Pil. Coloc. Co., Pil. Rhei Co.,
Ext. Hyosc., aa. gr. j.   M. ft. pil.

A very favourite remedy for constipation : acts mildly and may
be continued for ten days.

See F. 170, et seq.

(155.)

Pot. Iodid. et Pot. Bromid.

℞ Potass. Iodid., gr. jss. ; Potass. Bromid., gr. iijss. M. ft. pil.

A very convenient form for administering these valuable agents
when only small doses of Bromide are required.

Dose—Two or three twice or three times a day.

(156.)

Bismuth c. Ferro.

℞ Bismuth. Sub. Nit., gr. vij. ; Ferri Redacti, gr. ij. ; Ext.
Hyoscyami, gr. ij.   M. ft. pil. 2.

A very useful remedy in some forms of dyspepsia with anæmia.

Dose—One or two with food.

(157.)

Quinæ, Rhei, et Sodæ Carb.

℞ Quinæ Sulph., gr. ½ ; Ext. Rhei, gr. ½ ; Sodæ Carb.
Exsic., gr. ijss.   M. ft. pil.

A very useful remedy for the treatment of the diseases of childhood.

See also F. 50.

(158.)
Scillæ et Ipecac. et Stramonii.
(*Chest Hospital.*)

℞ Pil. Scillæ et Ipecac., gr. iv. ; Ext. Stramonii, gr. ½. M. ft. pil.

Gives speedy relief in asthma.

Dose—One only. It may, however, be repeated after an interval of some hours.

See Stramonium, Lupulin, and Henbane, F. 153.

(159.)
Pot. Bicarb. et Colchici.

℞ Potass. Bicarb., gr. v. ; Ext. Colchici, gr. ⅛; Ex. Aconit. Alcoholic., gr. $\frac{1}{18}$. M ft. pil.

This formula, it is believed, will be found a convenient and very efficacious remedy for acute attacks of rheumatism, gout, and other painful affections dependent on the lithic acid diathesis.

Dose—In acute attacks two for the first dose, one every hour for three hours, and then one every two, three, or four hours, according to the circumstances of the case.

(160.)
Zinci Ox. c. Morphiâ.

℞ Zinci Oxid., gr. ij. ; Morphiæ Hydroch., gr. $\frac{1}{12}$. M. ft. pil.

Oxide of Zinc is very useful in *chorea,* and in combination with Hemlock with Belladonna it is highly spoken of as a remedy for hooping-cough.

In chronic dysentery and in epilepsy, it is questionable if it has any advantage over the sulphate. *See F.* 31.

According to Dr. Marcet, it is the proper remedy for the nervous symptoms present *in chronic alcoholism.*

(161.)
Zinci Ox. et Hyoscyami.

℞ Zinci Oxid., gr. ijss. ; Ext. Hyoscy., gr. ij. M. ft. pil.

For the relief of night sweats in phthisis and other exhausting diseases, there are few remedies more useful than Oxide of Zinc. Hyoscyamus is substituted for the Morphia.

(162.)
Terebinth. et Rhei.

(Guy's Hospital.)

℞ Ol. Terebinth., gr. ijss.; Saponis Dur., gr. j.; Pulv. Rhei, gr. j. M. ft. pil.

(163.)
Pil. Alterativæ.

(London Hospital.)

℞ Pil. Coloc. et Hyosc., gr. iij. ; Pil. Hydrarg., gr. jss.; Pulv. Ipecac., gr. ¼ ; Ext. Colchici Acet., gr. ¼. M. ft. pil.

(164.)
Pil. Aperientes.

℞ P. Aloes Barb., gr. xxxvj. ; P. Jalapæ, gr. xxiv. ; P. Colocynth, gr. xij. ; Cambogiæ, gr. vij. ; Saponis Dur., gr. xij.; Ol. Carui, ♏ vj. Ft. pil. 24.

(164 A.)
Pil. Aper. c. Cal.

As above, with gr. ½ of Calomel in each pill.

The Extracts of Colocynth and Jalap of standard quality are very costly, and the above formula is devised to furnish an efficient aperient and antibilious pill, of good and cheap materials, for the common requirements of club and parish practice.

Dose—Two or three at bedtime.

See F. 174.

(165.)
Pil. Diaphoretic. Co.

℞ Camphoræ, gr. ½ ; Antim. Tart., gr. ₁/₁₈ ; Potassæ Nitratis Exsic., gr. ivss. Ft. pil.

Dose—For adults, one or two every hour for the first four hours, then every two hours for eight hours, and afterwards repeat every four hours ; for children from seven to twelve years of age, one every three or four hours, either whole or dissolved in warm gruel. It should be remembered that children are very sensitive to Antimony, and after a few doses it is generally desirable to substitute Ipecac.

See F. 98 and 102.

(166.)
Pil. Expectorans.
(London Hospital.)

℞ P. Ipecac., gr. j. ; P. Scillæ, gr. ¼ ; Ext. Hyoscyami, gr. jss. M. ft. pil.

Dose—One every three or four hours.

(167.)
Pulv. Antim. et P. Doveri.

℞ Pulv. Antimonialis, gr. ij. ; Pulv. Ipecac. c. Opio, gr. iij. M. ft. pil.

An active sudorific.

Dose—Two for the first dose, and one every four hours afterwards.
See F. 110.

(168.)
Aloes, Ferri, et Quinæ.

℞ Ferri Sulph. Ex., gr. ij.; Quinæ Sulph. gr. j. ; Ext. Aloes Aqu., gr. ½. M. ft. pil.

The action of Aloes is heightened by the Sulphate of Iron. This is a very useful combination in the *atonic* forms of constipation, with anæmia and debility. It is also successfully prescribed in amenorrhœa and deficient menstruation.

Dose—One, twice a day, with food.

(169.)
Aloes et Atropiæ.
(Dr. Macario; Pil. Antistyptic No. 1.*)*

℞ Ext. Aloes Soc., 5 centigrammes ; Ferri Sulph., 10 centigrammes ; Atropine, ⅓ of a milligramme.

In a communication to the *Lyon Médical*, Dr. Macario, of Nice, observes, that "in treating constipation most practitioners confine themselves to enemata, laxatives, or more or less irritating purgatives, which do more harm than good," and wishes to make known a remedy which he says may be truly termed "heroic," one which he has employed during twelve years with constant success, and therefore regards as infallible.

He refers particularly to two common forms of Constipation : 1st. *Nervous*, produced by intestinal excitement, with *deficient secretion*. 2nd. *Atonic*, produced by deficient contraction of the muscular coat of the intestine, which bad anti-hygienic habits have induced to keep up.

For nervous constipation he recommends the above pill.

In the atonic form, one centigramme of powdered *Nux Vomica* is substituted for the Atropine.

By the aid of these pills, regular stools may be procured, even in obstinate constipation, dependent on cerebral disturbance, paraplegia, etc. ·

Dr. Macario gives from one to three of these pills immediately after dinner, the object being to produce an easy, natural, non-diarrhœic evacuation. The use of these "*Antistyptic*" pills should not be continued indefinitely without an interval, it being of importance to allow the organs to resume their spontaneous action without any auxiliary.

(170.)
Aloes et Nucis Vomicæ.

(*Dr. Macario; Pil. Antistyptic No. 2.*)

The same as above, with 1 centigramme of powdered Nux Vomica in place of the Atropine.

For other remedies for habitual constipation, see F. 118, 131, 132, 168.

(171.)
Aloetic Granules.

Containing simply one grain of pure Aqueous Extract of Aloes, free from resin.

"Few medicines," says Dr. West, "act more mildly or more certainly in children than Aloes." The bitterness and bulkiness of the decoction and powder oppose great difficulty to its administration to young children. *Castor Oil, Senna,* and *Jalap,* for a like reason, and also on account of the nausea and griping which they occasion, are frequently inadmissible. While the habitual use of calomel and grey powder, simply to overcome constipation, is highly injurious. These granules, prepared with the finest aloes that can be obtained, will be found by far the most efficient and convenient aperient that can be devised for young children; they may be given at all ages, and are especially useful in the treatment of those obstinate forms of constipation which so often occur in infancy. In every case where a *mercurial* is not indicated, they may be employed with great advantage. Being perfectly *tasteless* there is no difficulty in giving them whole or in parts in a sweet preserve, to children who are too young to be taught to swallow them. In adults who suffer simply from sluggish bowels, one granule taken daily with dinner secures comfortable relief.

See Aloes and Iron . . . . . . F. 68, 169, 170.
,,   ,,   ,,   and Quinine . F. 132, 168, 134.
,,   ,,   ,,   and Nux. Vom. F. 118, 131.
And Glycæol of Aloes.

(172.)
Nux Vomica Granules.

These are prepared with the Alcoholic Extract, half a grain in each, in such a manner as to *ensure immediate solution and diffusion* over the mucous coat of the stomach.

Dose—One granule three times a day with meals.

Use—In constipation depending upon want of tone and imperfect propelling power of the colon, *Nux Vomica*, either alone or combined, is a very efficient remedy. Besides its action on the muscular coat of the bowel, it is a direct stimulant to the spinal cord, it acts directly on the mucous coat of the stomach; it increases appetite, promotes digestion, as well as aids the proper unloading of the bowels. In headaches, especially congestive headaches, morning sickness in pregnancy, in depression consequent upon over stimulation, it is very valuable. In gastralgia, pyrosis, chronic catarrh of the stomach, and dyspepsia, attended with coated tongue, flatulence, acidity, heartburn, the simple *Nux Vomica Granules* will afford considerable relief. One may be taken half an hour before meals, two or three times a day.

See Nux Vomica with Aloes . . . . F. 118, 131.
,,   ,,   ,,   and Iron . F. 170.
,,   ,,   with Quinine . . . F. 75.
,,   ,,   ,,   and Iron . F. 81.
,,   ,,   with Comp. Rhubarb . F. 54.
,,   ,,   ,,   Colocynth . F. 55.
,,   ,,   ,,   ,,   . F. 154.
For combinations with Phosphorus, see F. 189, et seq.

(173.)
Pil. Tonici c. Cinchonin.

R Ferri Sulph. Exsic., gr. jss.; Pulv. Nucis Vom., gr. j.; Cinchon. Sulph., gr. j.; Aloes Soc., gr. ½. M. ft. pil.

Dose—One or two, three times a day, with or after food.

A very useful and efficient tonic and excellent substitute for quinine, devised for parish and club practice. In many cases as effective as more costly medicines.

(174.)
Pil. Crotonis c. Hydrargyro.

℞ Ol. Crotonis, ʒss.; Calomel, gr. 144; Ext. Jalapæ, gr. 144; Ext. Coloc. Comp., gr. 360; Ext. Hyoscyami, gr. 72; Gingerinæ, gr. x. M. ft. pil. 144.

Dose—Two at bedtime.

A very efficient purgative and antibilious pill; may be prescribed when an *active* cathartic is indicated.

(175.)
Pil. Ferri Iodid. Comp.
(*Dr. Buckler, Baltimore.*)

℞ Potassii Iodid., gr. ij.; Ferri Iodid., gr. j.; Iodi, gr. ₁'₀.; Ext. Conii, gr. j. M. ft. pil.

Dose—One, three times a day, soon after food.

Useful in *scrofulous* and *strumous affections of the glands in cachectic subjects*. In constitutional syphilis affecting the bones and periosteum, *also in chronic rheumatic arthritis*.

(176.)
Pil. Camphoræ et Opii.
(*St. Mary's.*)

℞ Camphoræ, gr. ij.; Pulv. Opii, gr. j.

Dose—One at bedtime.

Useful in restlessness, with irritability of sexual organs. Also in venereal affections, especially if chordee or chancre be present, or nocturnal emissions. *See F.* 97.

(177.)
Pil. Camphoræ et Hyoscyami.
(*Fever.*)

℞ Camphoræ, gr. ij.; Ext. Hyoscyami, gr. iij. M. ft. pil.

Dose—As above.

In spasmodic affections of the uterus, bladder, and urethra, and as above, to which it is to be preferred when congestion of the liver or constipation is present. In these cases 4 grs. of blue pill may be administered with each dose of either preparation.

(178.)
Dr. Livingstone's Fever Pill.

℞ Jalapæ Res., P. Rhei, aa gr. vj.; Calomel, gr. iij.; Quinæ Sulph., gr. iij. M. ft. pil. iv.

(179.)
Quinæ et Ferri Lact. c. Ignat. Amar.

℞ Quinæ Sulph., gr. j.; Ext. Ignat. Amar., gr. ½; Ferri Lactatis, gr. ij.   M. ft. pil.

This is a very excellent combination of the Lactate of Iron with Quinine.   It may be used in cases where the Sulphate is not well borne.

Dose—One three times a day.

(180.)
Anthemidis Rhei c. Zingib.

℞ Ext. Anthemidis, gr. iij.; Ext. Rhei., gr. ij.; P. Zingib., gr. ij.   M. ft. pil., ij.

Dose—Two twice or one three times a day.

A useful vegetable tonic and stomachic in many forms of dyspepsia.

(181.)
Scillæ c. Colchici.
· (*Skin.*)

℞ Pil. Scillæ Co., gr. iij.; P. Colchici, gr. j.; P. Opii, gr. ¼.   M. ft. pil.

Dose—One twice a day in thoracic complications in gouty and rheumatic constitutions.

(182.)
Zinci Valer. Comp.
(*London.*) ·

℞ Zinci Valer, gr. ½; Quinæ Sulph., gr. ⅓; Pil. Rhei. Co., gr. j.; Ext. Gent., gr. ij.   M. ft. pil.

Dose—One or two twice or thrice daily.

A capital combination in many hysterical cases.

(183.)
Pil. Exect. c. Ammonio.

℞ Ammoniæ Carb., gr. iij.; Pil. Scillæ Comp., gr. iij.; Pulv. Lobeliæ, gr. j.; Pulv. Ipecac. c. Opio, gr. ijss.   M. ft. pil. 2.

A useful stimulating expectorant and antispasmodic, in chronic catarrh, bronchitis, asthma, etc.   Relieves dyspnœa from bronchial congestion.

Dose—One every three or four hours.

(184.)
Quinæ Arsenitis c. Ferro.

R Quinæ Arsenitis, gr. ⅛; Ferri Redacti, gr. iij.   M. ft. pil.

Useful in neuralgia, also in chronic skin diseases; a powerful hæmatinic and nerve tonic.

See Antineuralgic Pills, F. 122.

(185.)
Zinci Val., Camph., et Belladonnæ.

℞ Zinci Valer., gr. j.; Camphoræ, gr. ij.; Ext. Belladonnæ, gr. ⅓.

A soothing nervine tonic suitable in hysterical and epileptic cases where there is great irritability and sleeplessness.

Dose—One or two three times a day.

(186.)
Chloral c. Morphiâ et Cannabis Ind.

R Chloral Hydratis, gr. v.; Morphiæ Hydroch., gr. $\frac{1}{12}$; Ext. Cannabis. Ind., gr. ⅓.   M. ft. pil.

Dose—One or two at bedtime.

A convenient substitute for Chlorodyne.

(187.)
Hydrarg., Colchici, et Rhei Co.

R Pil. Hydrarg., gr. j.; Pil. Rhei Co., gr. iij.; Ext. Colchici, gr. ½; Ext. Hyoscyam, gr. ¼.   M. ft. pil.

A mild alterative, aperient, and diuretic medicine very useful in cases of rheumatism, gout, and their complications.

Dose—One or two every night at bedtime.

(188.)
Hyd. c. Rheo et Sodâ.

℞ Hyd. c. Cretâ, Soda Carb. Exsic., āā. gr. ij.; Ext. Rhei, gr. j.   M. ft. pil. 2.

Dose—One or two twice or three times a day.

An excellent alterative for children; given with benefit in liver and stomach disorders.

(189.)

Pil. Phosphori Mollis.* = 1 in 40.

$$5 \text{ grains} = \tfrac{1}{8} \text{ grain Phosphorus.}$$
$$2\tfrac{1}{2} \text{ ,,} = \tfrac{1}{16} \text{ ,,} \qquad \text{,,}$$
$$\cdot 2 \text{ ,,} = \tfrac{1}{20} \text{ ,,} \qquad \text{,,}$$
$$1\tfrac{1}{4} \text{ ,,} = \tfrac{1}{32} \text{ ,,} \qquad \text{,,}$$

Pearl coated Pills containing gr. $\tfrac{1}{32}$, $\tfrac{1}{20}$, $\tfrac{1}{16}$, and also the following combinations, may readily be obtained at all respectable dispensing establishments, and of Messrs. H. & T. Kirby & Co., 14, Newman Street.

(190.)

Pil. Phosphori et Nucis Vomicæ.

℞ Phosphori Pur., $\tfrac{1}{100}$ gr.; Ext. Nucis Vomicæ, $\tfrac{1}{4}$ gr. M. ft. pil.

Dose—One or two pilules three times a day, after meals.

PHOSPHORUS and NUX VOMICA, combined in the proportions above indicated, is a mild but valuable remedy. As a nutritive tonic and stimulant to the nervous system, especially the spinal cord, it is admirably adapted for the treatment of a large number of nervous disorders dependent on defective nutrition and debility, especially in the diseases of childhood. It increases appetite and promotes digestion. It may be safely given in all those cases in which the hypophosphites are employed with advantage.

(191.)

Pil. Phosphori c. Quinâ.

℞ Phosphori Pur., $\tfrac{1}{20}$ gr.; Quinæ Sulph., 1 gr. M. ft. pil.

Dose for Children from 7 to 10 years—One pilule three times a day. *For Adults*—Two twice or three times a day, at meals.

PHOSPHORUS and QUININE.—This is a valuable, and hitherto unattainable, combination of two powerful *restoratives—Phosphorus*, acting directly on the nervous system, nourishing the nervous centres and replacing the waste of nerve tissue, gives vigour to the functions of the brain and spinal cord, and *Quinine* giving tone to the digestive organs, and strengthening and improving the condition

---

* So called to distinguish it from the Pil. Phosphori, recently introduced among the "*additions*" to the *B.P.* New edition. 1874. When ordered in prescriptions the name "Kirby" should be added to ensure the right preparation being obtained. See pamphlet *On the Administration of Phosphorus as a Remedy for Loss of Nerve Power, &c.* By Dr. E. A. Kirby. Price 1s. H. K. Lewis, 136, Gower Street, London.

of the whole system. As a stimulant and nutritive tonic, it is one of the best that can be devised. There are very few cases where Quinine is indicated in which this combination may not be prescribed with the best effect.

(192.)

Pil. Phosphori c. Quinâ, et Nucis Vomicæ.*

℞ Phosphori Pur., $\frac{1}{60}$ gr.; Quinæ Sulph., gr. j.; Ext. Nucis Vomic., $\frac{1}{4}$ gr. M. ft. pil.

Dose—One pilule three times a day. *For Adults*—Two twice or three times a day, with food.

Where Iron is not required this is the best formula that can be selected for *neuralgia*. In severe cases two may be taken every four hours.

For a similar combination with Iron, see F. 197.

(193.)

Pil. Phosphori c. Ferro et Nucis Vom.*

℞ Phosphori Pur., gr. $\frac{1}{33}$; Ferri Redacti, gr. iij.; Ext. Nucis Vom., gr. $\frac{1}{4}$. M. ft. pil.

Dose—One or two pilules three times a day with food.

Ferrum Redactum is a powerful hæmatinic even in small doses, and is well adapted to promote the blood-restoring properties of the metal. The absence of astringency renders it peculiarly useful in the treatment of diseases depending on poverty of blood, in which other preparations of iron would not be admissible. It is more than probable that in every case where iron is wanting phosphorus is also deficient, and it is only reasonable to believe that where the red globules are reduced one-third or one-half, and the liquor sanguinis is poor in albumen (which is commonly the case in *anæmic* conditions), the normal proportion of phosphorus is of necessity much below the healthy standard. The combination, therefore, is a highly useful one.

(194.)

Pil. Phosphori et Nucis Vom. Fort.

℞ Phosphori Pur., $\frac{1}{33}$ gr.; Ext. Nucis Vomic., $\frac{1}{4}$ gr. M. ft. pil.

Dose—One or two three times a day with food.

This combination is found exceedingly useful in *atonic dyspepsia, lowness of spirits*, and in that general condition of depression and loss of power popularly known as *below par*, and in *break down* from

---

* Mr. Jabez Hogg suggests that these combinations, F. 192 and F. 193, promise to be useful in *atrophy of the optic nerve*.

overwork and mental fatigue. PHOSPHORUS and NUX VOMICA are probably the only *medicines* which can be relied upon as *sexual stimulants* or possess real aphrodisiac power. Their administration, however, with this view requires caution : large doses are neither necessary or desirable nor should they be long continued without intermission ; one or two pills twice or three times a day, according to the circumstances of the case, may safely be prescribed for two or three weeks in succession with advantage. In cases of *impotence occurring in old and debilitated subjects* gr. $\frac{1}{15}$, or even gr. $\frac{1}{15}$, of Phosphorus, either alone or in combination with Quinine and Nux Vomica, may be administered with the best results. But after success has been attained, the remedy should be left off for a time.[*]

(195.)
Pil. Phosphori c. Ferro.

℞ Phosphori Pur., $\frac{1}{30}$ gr.; Ferri Redacti, 3 gr. M. ft. pil.

Phosphorus and Iron is a powerful nutritive tonic and blood restorer. It is especially valuable in TUBERCULAR DISEASES, *consumption, tabes mesenterica, scrofula*, and the strumous diseases and cachetic conditions of children. It is given with great advantage in *anæmia, chlorosis*, in *sciatica* and other neuralgic affections; also in furuncular inflammations, carbuncles, boils, etc.

Dose—For adults, one or two twice or three times a day with food. For children between 7 and 12 years of age, one twice or thrice daily with food.

See F. 163.

(196.)
Pil. Phosphori c. Ferro et Quinâ.

℞ Phosphori Pur., $\frac{1}{30}$ gr.; Ferri Redacti, 3 gr.; Quinæ Sulph., $\frac{1}{2}$ gr. M. ft. pil.

The uses of Iron and Quinine in combination are too well known to need any remark. The addition of Phosphorus intensifies their action, while it imparts additional power by supplying nutrition to the nervous system.

Dose as above.

(197.)
Pil. Phosphori Comp.

℞ Phosphori Pur., $\frac{1}{30}$ gr.; Ferri Redacti, 3 gr.; Quinæ Sulph., $\frac{1}{2}$ gr.; Strychniæ, $\frac{1}{30}$ gr. M. ft. pil.

[*] See ACTON'S *Functions and Disorders of the Reproductive Organs*, page 85.

E

This is a valuable and highly efficient combination. It will be found most efficacious in that numerous class of cases characterized by impoverished blood and loss of nerve power, and in which Iron and Phosphorus are both strongly indicated. It must be obvious that this preparation admirably fulfils the purposes for which these powerful therapeutic agents are prescribed. It will be found more effectual than the hypophosphites in syrups, and other feeble modes of exhibiting Phosphorus.

It is unnecessary to particularize every disease in which this formula is useful. There can be no doubt that these powerful remedies act *together* in a *special* manner, and answer different indications more fully than when administered separately.

In many cases this form may be substituted for F. 194 or 192.

Dose—One or two twice or three times a day with food.

(198.)
Pil. Phosphori c. Morphiâ.

℞ Phosphori, gr. $\frac{1}{15}$; Morphiæ Hydroch., gr. $\frac{1}{12}$; Zinci Valer., gr. j. M. ft. pil.

In phthisis when accompanied with hysterical irritability and troublesome cough with little febrile disturbance it both soothes and supports.

This and the two following preparations are suggested as being well adapted for the treatment of phthisis. They are recommended in all those stages in which the Hypophosphites* have proved useful, as far more efficient and reliable remedies. They may be given advantageously during a course of cod-liver oil. I have seen cases in which marked improvement has resulted. Formulæ 193 and 195 are also frequently prescribed for this purpose.

Dose—One, twice'or thrice daily, or two at bedtime.

(199).
Pil. Phosphori c. Cannabis Ind.

℞ Phosphori Pur., gr. $\frac{1}{15}$; Ext. Cannabis Ind., gr. $\frac{1}{4}$.

As above, when Morphia is contra-indicated, and to produce sleep.

A good aphrodisiac in some cases where the combination with Nux Vomica fails.

Dose—One or two twice or three times a day.

---

* See DR. J. F. CHURCHILL, *Ranking's Abstract*, vol. xxvi., page 41.

(200.)

Pil. Phosphori et Aconiti.

℞ Phosphor. Pur., gr. 1/30; Ext. Aconiti Alc., gr. 1/18. M. ft. pil.

This combination has been suggested by Dr. Prosser James. It will be found very useful in the treatment of phthisis with pyrexia.

Dose—One every four hours.

MISTURÆ.

The following formulæ are so framed that, with a trifling modification, the Mixtures may be CONCENTRATED, and in this form *kept ready* for use, a matter of interest to those who desire to curtail the labour of dispensing. They correspond to Mixtures in constant use at the London Hospitals, and will be found to supply the ordinary requirements of practice. In many cases they will be found useful *as bases*, additions being made to them to suit individual wants.

Mistura Magnes. Sulph. Acida.

℞ Magnes. Sulph , ʒvi. ; Acid. Sulph. Dil., ʒjss. ; Aquæ, ad ℥vj.

A simple saline aperient.

Mistura Magnesiæ c. Magnes. Sulph. Alkalina.

(*Syns.: Mist. Magnes. c. Colch., Mist. Alba.*)

℞ Magnes. Sulph, ʒvj. ; Magnes. Carb. Pond., ʒj. ; Vin. Colchici, ʒj. ; Tr. Aconiti, ♏xij. ; Glycerinæ, ʒj. ; Aquæ, ad ℥ vj.

To relieve portal congestion. A good purgative in gout, rheumatism, etc.

Refrigerant, cathartic, and antacid ; slightly depressant if pushed.

Mistura Magnes. Sulph: c. Rosâ.

(*Mist. Rosæ Aperiens.*)

℞ Magnes. Sulph., ʒvj. ; Acid. Sulph. Arom., ʒj.; Tr. Zingib., ♍xxiij.; Glycerinæ, ♍xxij. ; Infusi Rosæ, ad ʒvj.

This combination is an improvement on the "Red Mixture" of the hospitals. It is an elegant and convenient mode of administering a bulky and nauseous medicine. As a refrigerant cathartic, Sulphate of Magnesia is perhaps more generally employed than any other of its class. It operates mildly yet effectually, both by augmenting the secretions and by increasing peristaltic action; and there are but few cases in which it is indicated that this mixture may not be given with benefit. Generally, it is useful in all febrile, congestive, and inflammatory affections. Particularly it is beneficial in hæmorrhage during and after abortion, hæmoptysis, menorrhagia, and epistaxis. Gallic Acid (gr. x. doses) may be added in severe cases.

Mistura Acidi Sulph. Aromatica.

℞ Acid. Sulph. Arom., ʒj. ; Tr. Aurantii, ʒij.; Tr. Cardam. Co., ʒij. ; Sp. Chloroformi, ʒjss. ; Aquæ, ad ʒvj.

A good diarrhœa mixture.

Mistura Acidi Nitrohydrochlorici.

℞ Acid. Nitrohydrochlorici Dil., ʒjss.; Syrupi, ʒiij.; Sp. Myristicæ, ʒj. ; Infusi Quassiæ, ad ʒvj.

Mistura Acidi Phosphorici c. Ferro.

℞ Acid. Phosph. Dil., ʒij.; Tr. Ferri Perchlor., ʒjss.; Sp. Chloroformi, ʒij. ; Infusi Quassiæ, ad ʒ vj.

A favourite remedy of the late Dr. Hodgkin.

Mistura Acidi Nitrohydrochlorici c. Ferro et Strychniâ.

(*Mist. Tonici.*)

℞ Tr. Ferri Perchlor., ʒij. ; Acid. Phosph. Dil., ʒjss.; Acid. Nitrohydrochlor. Dil., ʒj.; Liq. Strychniæ, B.P., ʒss. ; Sp. Chloroformi, ʒj. ; Glycerinæ, ʒj.; Infusi Quassiæ, ad ʒvj.

I prescribed this tonic at the City Dispensary extensively, and found it invaluable in debility and nervous depression.

Mistura Alkalina (Potash) c. Gentianâ.

℞ Potassæ Bicarb., ʒjss.; Syrupi, ʒij.; Sp. Ammon. Arom.,
ʒjss; Infusi Gent. Co., ad ℥vj.

Mistura Alkalina (Soda) c. Calumbâ.

℞ Sodæ Bicarb., ʒjss.; Sp. Chloroformi, ʒjss.; Infusi Calumbæ, ad ℥vj.

Mistura Alkalina Aromatica.

℞ Sp. Ammon. Arom., ʒij.; Sp. Chloroformi, ʒij.; Infus.
Aurant. Co., ad ℥vj.

Mistura Alkalina Aromatica c. Rheo.
(*Mistura Stomachica.*)

℞ Infusi Rhei, ℥jss.; Sp. Ammon. Arom., ʒjss.; Infus.
Gentianæ Co., ad ℥vj.

A gentle laxative, tonic, antacid, and stomachic medicine, and as
such it is applicable to the treatment of a large number of cases met
with in everyday practice. *Carbonate of Soda* or *Bicarbonate of
Potash* may be added when their employment is indicated. With
the former it makes the well-known *brown mixture*, Soda c. Rheo, of
the hospitals. Useful in many forms of dyspepsia, especially so
when occurring in cachectic subjects and weakly children, and with an
increased dose of Ammonia, in disorders which follow upon excesses in eating and drinking—a numerous class of disorders.

Mistura Ammoniæ Acet.
(*Mist. Salinæ.*)

℞ Liq. Ammon. Acet., ℥jss.; Sp. Ether. Nit., ʒij.; Syrupi
Croci, ʒss.; Mist. Camphoræ, ad ℥vj.

Mistura Ammoniæ Acet. Composita.
(*Mist. Salinæ Co.   Mist. Diaphoretica.*)

℞ Liq. Ammon. Acet., ℥jss.; Vin. Antim. Tart., ♏lxxij.;
Sp. Ether. Nit., ʒj.; Tr. Aconiti, ♏vj.; Sp. Camphoræ, ʒj.;
Glycerinæ, ʒjss.; Aquæ, ad ℥vj.

The mixture is applicable to the treatment of nearly all acute
febrile and inflammatory conditions. It will be serviceable in every
case where it is desired to increase the cutaneous exhalation without
excitation or depression. When it is desired to secure a depressant action, the dose of Antimony should be increased either by adding more

Antimony to the Mixture, or by prescribing it in pills of the required strength to be taken with each dose. When on the contrary, in cases of feeble circulation, with cold skin, it is desired to stimulate the heart's action and to excite the nervous system, pills containing Hydrochlorate or Carbonate of Ammonia may most advantageously supplement each dose of the mixture. To produce *sweating* it may be desirable to prescribe simple Dover's Powder (Glycecols or Pilules) with the mixture, or F. 98 or 100. F. 102 containing Ipecacuanha adds an expectorant effect if that be desired.

Mistura Ammoniæ Effervescens.

℞　Ammoniæ Carb., ℨij.; Syrupi, ℥ss.; Aquæ, ad ℥vj.; Sp. Ammon. Arom., ℨj.; Tr. Aurantii, ℨj.

One fluid ounce is neutralized by xxij. grs. of Citric Acid or one large tablespoonful of Lemon Juice.

Mistura Ammoniæ c. Senegâ.

℞　Sp. Chloroformi, ℨjss.; Tr. Scillæ, f ℨjss.; Glycerinæ, f ℨij.; Ammon. Carb., gr. xxiv.; Infus. Senegæ, ad ℥vj.

A very useful stimulating expectorant.

Mistura Ammoniaci, Ipecacuanhæ, et Lo-beliæ.

℞　Misturæ Ammoniaci, ℥iij.; Vini Ipecac., ℨj.; Tr. Lobeliæ, ℨj.; Glycerinæ, ℥ss.; Aquæ, ad ℥vj.

In chronic bronchitis and asthma.

Mistura Astringens c. Hæmatoxyli.

℞　Acid. Sulph. Arom., f ℨjss.; Tinct. Cardam. Co., f ℨjss; Sp. Chloroformi, f ℨij.; Tr. Opii, f ℨss.; Ext. Hæmatoxyli, ℨij.; Aquæ, ad ℥vj.

A very useful diarrhœa mixture.

Mistura Cascarillæ Composita.

(*Mist. Tussi.*)

℞　Tr. Camph. Co., ℨiij.; Acid. Nitrici Dil., ℨj.; Vini Ipecac., ℨj.; Glycerinæ, ℥ss.; Tr. Scillæ, ℨss.; Infusi Cascarillæ, ad ℥vj.

A useful cough mixture.

Mistura Chiratæ Composita.

℞　Acid. Nitrohydrochlorici Dil., ℨjss; Glycerinæ, ℨjss.; Infus. Chiratæ, ℥ ij.; Infusi Cinchonæ Flav., ad ℥vj.

Mistura Cinchonæ Acida.

(*Consumption Hospital.*)

℞ Acidi Nitrici Dil., ʒjss. ; Glycerinæ, ʒij. ; Infus. Cinchonæ Flav., ad f ʒvj.

Useful in debility connected with the alkaline and phosphatic diathesis, in syphilis and secondary syphilitic eruptions of the skin, and atonic dyspepsia. In the advanced stages of hooping-cough it is particularly useful; also in atonic diarrhœa and in the low stages of fevers, especially so in typhoid and scarlatina. This is a most useful medicine to keep ready. It is also an excellent gargle in affections of the throat attended with ulceration, etc.

Mistura Cinchonæ Ammon. et Chloroformi.

℞ Ammon. Carb., gr. xviij.; Sp. Chloroformi, ʒjss.; Tr. Cinchonæ Co., ʒiij. ; Glycerine, ʒjss. ; Decoct. Cinchonæ, ad ʒvj.

Stimulant, tonic, and restorative. Useful generally in asthenic conditions, the advanced stages of febrile and inflammatory diseases, erysipelatous inflammations, etc.

Misturæ Cinchoniæ.

(*University College Hospital.*)

℞ Cinchoniæ Hydrochlor., gr. xij. ; Acid. Hydrochlor. Dil., ♏xij. ; Aquæ, ad ʒvi.

A capital tonic in dyspepsia, and useful substitute for quinine in many cases.

Mistura Copaibæ Composita.*

(*London Hospital*).

℞ Bals. Copaibæ, ♏xv. ; Liquor Potassæ, ♏xv. ; Ol. Cubebæ, ♏x. ; Sp. Ether. Nit., ʒss. ; Aquæ Camphoræ, ad ʒj.

Mistura Diuretica.

℞ Potassæ Acetatis, ʒij. ; Sp. Ether. Nit., ʒiij.; Aceti Scillæ, ʒij. ; Succi Scoparii, ʒvj. ; Aquæ, ad ʒvj.

Mistura Ergotæ Ammoniata.

(*University College Hospital*).

℞ Ammon. Carb., gr. xxiv. ; Liquor Ergotæ, ʒij.; Tr. Lavand. Co., ʒjss. ; Sp. Chloroformi, ʒjss.; Aquæ Camphoræ, ad ʒvj.

---

* MESSRS. KIRBY & Co. (14, Newman Street) prepare an admirable solution of Copaiba, readily miscible with water—a preparation which saves much trouble in dispensing.

Mistura Hydrargyri, Iodidi, et Sarsæ.
(*University College Hospital.*)

℞ Hydrargyri Perchloridi, gr. ss. ; Potassii Iodidi, gr. xxxij. ;
    Decoct. Sarsæ Co., ad ℥viij.

The Hospital orders water in this mixture.

Dose—℥j. to ℥jss.

Mistura Potassii Iodidi Composita.
(*Skin Hospital.*)

℞ Iodi., gr. iij. ; Liq. Potassæ Arsenitis, ♏ xxiv. ; Liq. Potassæ,
    ♏ xxiv. ; Tr. Cardam. Co., ♏ xxiv. ; Aquæ, ad ℥j.

Dose—℈j. ad ℈ij. in water, in cachectic, squamous, pustular, and
vesicular affections.

Mistura Potassii Bromidi Composita.
(*University College Hospital.*)

℞ Potassii Bromidi, gr. x. ; Sp. Chloroformi, ♏xviij. ; Infus.
    Quassiæ, ℥j.

Dose—℥ss. to ℥jss.

MIXTURES FOR CHILDREN.

(For other Remedies for Children, see Glycecols, Syrups, and Pilulæ.)

Mistura Carminativa Antacid.

℞ Sodæ Bicarb., gr. xij. ; Glycerinæ, f ℈j. : Infus. Aromat.
    Comp.,* ad ℥j.   M. ft. M.

This is a very nice carminative medicine for children. It relieves
wind and griping, and it forms a good base for diarrhœa mixtures.

Dose—A teaspoonful ; for very young infants it may be diluted to
half strength.

Mistura Carminativa Aperiens.

℞ Potassæ Tartratis, ℥j. ; Liq. Sennæ Dulc., f ℥ij. ; Infus.
    Aromat. Comp., ad ℥iv.   M. ft. M.

A mild, cooling, saline aperient, palatable, and very efficient.
Useful especially in acid conditions of the prima viæ.

---

* **Infusum Aromatica Comp.**

℞ Cort. Cinnam., ℈ij. ; Sem. Cardam., ℈j. ; Rad. Zingib, ℈j. ; Caryophylli,
℈iv. ; Carui, ℈ij. ; Aquæ, ad Oj.   M. ft. s. a.

Mistura Ipecacuanhæ.

℞ Vini Ipecac., ♏xx. ; Succi Belladonnæ, ♏viij. ; Potassæ Nit., gr. xvj. ; Glycerinæ, f ʒj. ; Aquæ Amygdalæ, ad ʒj. M. ft. M.

Dose—ʒj. to ʒij.

Mistura Astringens.

℞ Ext. Hæmatoxyli, ʒj. ; Infus. Aromat. Comp., ʒj. M. ft. M.

In choleraic disease of infants, and in serous diarrhœa, dilute Sulphuric Acid may be added to this mixture with great advantage.

Dose—ʒj. to ʒij.

———

PULVERES.

(For formulæ for Powders for Children, see Glycecols.)

Pulvis Astringens.

℞ Pulv. Catechu Co., ʒij. ; Pulv. Aromat., ʒij. ; Cretæ Præp., ʒij. ; Sacchari, ʒj. ; Ol. Cinnamomi, ♏iv. M. ft. pulv.

For summer diarrhœa, no better remedy can be employed. It contains no Opium, and may therefore be prescribed without hesitation to children of all ages. Six drachms of this powder in six ounces of water forms at once a useful and safe "Diarrhœa Mixture." I have prescribed this mixture extensively, and have never been disappointed when given in suitable cases.

Dose—ʒj. to ʒij. ; in water, or brandy and water.

Pulvis Salinæ Effervescens.

℞ Sodæ Pot. Tart., ʒij. ; Sodæ Bicarb., ʒij. ; Acid. Tartarici, gr. cv. ; Ol. Limonis, gtt. iij. M. ft. pulv.

Dose—ʒss. ad ʒj. in a wine-glass of water is a febrifuge. As an aperient, ʒij. or ʒiij. in half a pint of water.

This powder properly prepared—*secundum artem,* keeps well and effervesces freely, evolving carbonic acid gas. It is a convenient ready remedy, and quickly dispensed. An agreeable "Fever Mixture."

ELIXIRS.

A popular form of medicine largely employed in the United States, being exceedingly pleasant to the taste, and at the same time possessing considerable medicinal value. The following are selected from formulæ recommended by the committee of the College of Pharmacy of Philadelphia.

The simple elixir contained in the following formulæ is a preparation containing cinnamon, orange, and other flavouring ingredients, which being sweetened, effectually disguises the taste of the medicines. They contain a *good portion* of *Alcohol*, which may account in a measure for their popularity.

Elixir of Yellow Peruvian Bark.

℞ Tinct. Cinchonæ Flav. Conc., ʒxxij.; Elixir Simplicis, q. s. ft. f ℥xvj. M. ft. Elixir.

This is an exceedingly pleasant and energetic tonic, febrifuge, and restorative. Each fluid dram contains the active principles of 10 grains of Bark.

Elixir Bark and Iron.

℞ Ferri Ammon. Cit., 128 grs.; Aquæ Dest., ℥j.; Elixir Cinchonæ Flav., ℥xv. M. ft. Elixir.

This valuable preparation excites languid appetite, gives zest to food, improves digestion, increases the strength, removes the pallor of debility, and gives firmness and precision to the action of the nervous system, with power to endure fatigue and resist disease. One fluid dram contains one grain of Iron Salt and the active principles of nearly 10 grains of Bark.

Elixir Pepsin.

℞ Pepsinæ, 256 grs.; Vini Xerici, ℥xiv.; Syrupi, ℥ij.; Ext. Zingib. Fluid., gtt. xxv. M. ft. Elixir.

This contains the active principle of the gastric juice in solution, forming an agreeable and elegant preparation. One fluid dram contains 2 grains of pure Pepsina Porci.

Elixir Bismuth.

℞ Bismuthi Amm. Citratis, 256 grs.; Aquæ Destillatæ, ℥j.; Elixir Simplicis, ℥xv. M. ft. Elixir.

This agreeable elixir contains 2 grains of soluble Citrate of Bismuth in one fluid dram, and is highly efficient in many painful affections of the stomach and bowels, being more active in smaller doses than the insoluble salts.

Elixir Pyrophosphate Iron.

℞ Ferri Pyrophosph., 128 grs.; Aquæ Dest., ℥j.; Elixir Simplicis, ℥xv. M. ft. Elixir.

The freedom from all unpleasant taste, and the ease with which this preparation is borne by even the most delicate, together with its ready assimilation with the food, and consequent rapid absorption, render this preparation specially valuable. It is used with benefit in those cases where a nervous tonic is indicated. One fluid dram contains 1 grain of Iron Salt.

Elixir Iron and Gentian.

℞ Ext. Gentianæ, 128 grs.; Ferri et Ammoniæ Cit., 128 grs.; Aquæ Dest., ℥j.; Elixir Simplicis, ℥xv. M. ft. Elixir.

In this elixir the valuable tonic properties of Gentian in combination with Iron form one of the most agreeable and effectual preparations extant. One fluid dram contains 1 grain of Iron Salt.

———

TROCHISCI.

The following formulæ merit particular attention. They are found to be exceedingly useful remedies, and are largely employed at the Throat Hospital.

Trochisci Acidi Benzoici. T.H.

Benzoic Acid, in powder, 175 grains in 350 = ½ gr. in each.

Dose—One lozenge every four hours; if used as a *voice* lozenge, one should be taken a quarter of an hour before using the voice.

Use—A most valuable stimulant and "voice lozenge" in cases of nervo-muscular weakness of the throat.

Trochisci Acidi Carbolici. T.H.

Pure Carbolic Acid, 350 grains in 350 = 1 gr. in each.

Dose—One lozenge four or five times daily.

Use—Antiseptic and stimulant.

Trochisci Acidi Tannici. T.H. 1¼ gr. each.

Dose—One lozenge every three or four hours.

Use—Strongly astringent.

The officinal preparation contains gr. ½ of the acid in each lozenge.

Trochisci Catechu. T.H.

Extract of Catechu, 700 grains in 350 = 2 gr. in each.

Dose—One lozenge every three hours.

Use—Astringent, but less powerful than the Tannic Acid.

B.P., gr. 1 in each.

Trochisci Cubebæ. T.H.

Pulv. Cubebæ, 200 grains in 400 = ½ gr. in each.

Dose—One every three or four hours.

Use—Very serviceable in diminishing excessive secretions of mucus from pharynx, larynx, or trachea. An improvement on " Brown Bronchial Troches."

Trochisci Guaiaci. T.H.

Contain of Guaiacum, 2 gr. in each.

Dose—One lozenge every two hours in acute inflammations, three times a day in chronic affections.

Use—A specific for arresting crescent inflammation of the tonsils, and useful in acute and subacute inflammation of the pharynx, and in acute follicular disease of the tonsils, etc., etc.

Trochisci Kino. T.H.

Contains pure extract of Kino, 2 gr. in each.

Dose—One lozenge every three or four hours.

Use—Astringent; rather less powerful than rhatany, and never likely to constipate.

Trochisci Krameriæ. T.H.

Contains pure extract of Rhatany, 3 gr. in each.

Dose—One lozenge every three or four hours.

Use—A very useful astringent. Rhatany is not so likely to disagree with the stomach as kino, catechu, and other astringents, nor does it often produce constipation.

Trochisci Morphiæ. B.P.

Each lozenge contains $\frac{1}{36}$ gr. Hydrochlorate of Morphia.

Dose—One lozenge every three or four hours.

Use—Sedative, for irritative coughs and painful conditions of the pharynx.

Troch. Morph. c. Ipecac., B.P.

Each lozenge contains $\frac{1}{36}$ gr. Hydrochlorate of Morphia, and $\frac{1}{12}$ gr. Ipecacuanha. *See Pilulæ, F.* 96.

Trochisci Potassæ Chloratis. T.H.

Chlorate of Potash, in powder, 1050 grains 350 = 3 gr. in each.

Dose—One lozenge every three or four hours.

Use—Stimulant and antiseptic. Useful in thrush and aphthous ulcerations, and as a general tonic to the mucous surface.

B.P., gr. 5 in each.

The above, together with many other formulæ, will be found in the list of Glycecols, and in that form they are certainly much more efficient.

TINCTURA.

The following and many of the Pharmacopœia Tinctures are conveniently dispensed in the manner already suggested.

Tinctura Cardamomi c. Quiniâ Conc.

SYN. *Liq. Quinæ Acida.*

℞ Chloroformi, 200 minims; Quinæ Sulph., 400 grs.; Acidi Sulph. Aromat., q.s. ; Tinct. Card. Co. Conc., ad Oj. M. ft. Tinct.

Four fluid drachms of this Tincture contains 12 grains of Sulphate of Quinine, and forms an elegant mixture when diluted with f ʒvss. of water.

Tinctura Chloroformi c. Opio.

℞ Tr. Opii, Sp. Camphoræ, Tr. Capsici, āā f ʒj.; Chloroformi, f ʒiij.; Sp. Vini Rect., ad f ʒv. M. ft. Tinct.

Each fluid dram contains about 100 *drops*, consisting of 12 minims of each of the first three ingredients, and 4½ *minims*, or 18 drops of Chloroform.

For popular 'use, in time of epidemic cholera, this preparation should be diluted one third.

See Tinct. Opii Ætherea.

Tinctura Colchici Etherea (American Form).

℞ Colchici, ʒvj.; Sp. Ether. Nit., Oj. vel q.s. Treat by displacement till Oj. of the Tincture is obtained.

Dose—20 to 30 drops.

This and the following preparation are used jointly for rheumatic and neuralgic symptoms. They substitute with advantage the alcoholic Tinctures of the same drugs.

Tinctura Guaiaci Etherea (American Form).

℞ Guaiaci Resinæ, ʒiij.; Sp. Ether. Nit., Oj. vel q.s. Treat by displacement until Oj. of the Tincture is obtained.

Guaiacum has long been a favourite remedy as a stimulating diaphoretic in chronic rheumatism, "cold" rheumatic pains, sciatica, etc., occurring in persons advanced in life. This preparation unites with it a diuretic action, and it is in some cases to be preferred to the Ammoniated Tincture. In dysmenorrhœa, amenorrhœa, and other uterine affections, it has been found very useful.

Dose—ʒss. to ʒij. diluted.

Tinctura Opii Etherea (American Form).

(*Asiatic Cholera Mixture*).

℞ Opii, Camphoræ, āā ʒj.; Ol. Caryophylli, f ʒj.; Capsici, ʒj.; Sp. Ether. Co. (Hoffmann's Anodyne), Oj. M. ft. Tinct.

Adult Dose—20 to 60 drops every third or fourth hour, according to circumstances. The diffusible character of the Ether is admirably adapted to heighten the effects of the important remedies it contains.

This preparation is a new remedy in this country, but it has been extensively employed in America. The formula is sufficient to indicate its value and the class of cases for which it is specially suited. In cases where it is desired to obtain an *immediate* effect, it is invaluable. When cholera is epidemic every practitioner will do well to arm himself with this remedy. It will enable him to arrest premonitory diarrhœa directly he is called in, and in that manner prevent a state of collapse, in which medicine appears so powerless.

Tinctura Veratri Viridis. U. S. P.

℞ Veratri Viridis, ʒviij.; percolate with Sp. Vini Rect. to f℥xvj. Tincture.

Cardiac Depressant and Sedative, used "to control the vascular system in cases of inflammatory diseases, especially rheumatic fever and gout. Depression and slowness of pulse appear to be characteristic symptoms of its action."—Dr. GARROD.

It is but little used in this country, but is highly esteemed in America in the treatment of pneumonia. It is said to be capable of rendering the pulse as low as thirty-five beats in a minute. I believe it to be largely employed in homœopathic practice, and with great success. It is not poisonous in the degree that Aconite is, and may therefore be more fearlessly used.

Dose—♏ v. ad ♏. xv.

N.B.—This tincture is more than double the strength of that of the British Pharmacopœia.

SYRUPUS.

The following are very efficient and agreeable medicines for children.

Syrupus Ipecacuanhæ. U. S. P.

℞ Ext. Ipecacuanhæ Fluidi, f ʒj.; Syrupi, f ℥xv. M. ft. Syr. f ℥j. = 3⅔ grs. Pulv. Ipecac.

Syrupus Krameriæ. U. S. P.

℞ Ext. Krameriæ, ʒij. (troy); Sacchari, ℥xxx. (troy); Aquæ, f ℥xvj. M. ft. Syr. f ℥j. = about 3 grs. Extract Krameria.

Syrupus Pectoralis (Linctus pro Tussi).

℞ Mellis, ʒij.; Syrupi Rhœados, ℥iij.; Liq. Morph. Bimec., ʒij.; Acidi Nitrici Dil., ʒiv.; Glycerinæ, ℥j.; Mucilaginis, ad ℥viij. M. ft. Linctus.

Dose—Adult ʒj. to ʒij. A favourite cough medicine.

Syrupus Rhei Aromat. (Spiced Syr. Rhubarb).

℞ Rhei, ʒijss.; Caryophylli Cinnamomi, āā ℥ss. Myristicæ, ʒij.; Diluted Alcohol, q.s. Treat by displacement until f ℥xvj. of tincture are obtained, and mix with Ovj. Syrup.

An excellent remedy for some forms of diarrhœa occurring in infancy.

Dose.—f ʒj. to f ʒij.

Syrupus Senegæ.

℞ Rad. Senegæ Contus., ʒiv. (troy); Alcohol Dilut., Oij. Treat by displacement, and evaporate the tincture obtained in a water bath at a temperature not exceeding 160° F. to f ʒviij. Filter, add sugar, ʒxv. (troy); dissolve by the aid of a gentle heat and strain.

Dose—For adults ʒj. to ʒij.

GLYCECOLS.

This is a new and highly convenient vehicle for administering medicines, in the form of a troche or lozenge, but differing widely from it in its physical characters, composition, and mode of manufacture. The basis is a compound of Glycerine and Isinglass,* Glycecolloid, possessing powerful solvent and antiseptic properties admirably adapted for the preservation and administration of medicinal substances.

The remarkable antiseptic and solvent power of the glycerine, together with its bland taste, its unchangeableness, and its neutral relation to the animal tissues, render it the most valuable excipient that can be used, especially for those medicines which it dissolves more readily than water, and gives to this form of medicine a utility and importance which it hitherto has not possessed.

Troches so prepared I propose to call *Glycecols*, in order to distinguish them from the lozenge of the confectioner, and I claim for them not only all the advantages which belong to that form of medicine as a means of local medication, and as an agreeable mode of administering certain medicines to children, but also the additional value derived from the properties of the Glycerine, as a therapeutic agent as well as an excipient.

The officinal troche is only *nominally* a pharmaceutical preparation, and its neglect as a form of medicine is doubtless due to the fact that its manufacture is in the hands of the confectioner; for although formulæ for their preparation appear in the Pharmacopœia, the pharmacist has no uniform and recognised practical method nor convenient mechanical appliances for dispensing medicines in this form with that expedition, accuracy, and neatness which pertain to pills, mixtures, or powders. Moreover the bases employed, viz. :— mixtures of gum and sugar, fruit paste, etc., are only suitable

---

* For some purposes the *Glycecolloid* is made with fine Gelatine prepared from calves' feet only.

excipients for comparatively few medicines, and being identical with those used in making the ordinary lozenge of commerce, the character of the medicated lozenge in no way differs from those sold in the shops as sweetmeats. These are grave objections and positive hindrances to their general employment in medicine, and it cannot be a matter of surprise that they are seldom prescribed by physicians, and for obvious reasons never used in general practice.

Glycecols are vastly superior to lozenges inasmuch as the utility of the latter is, at best, very limited in its application, whereas the Glycecol possesses properties which peculiarly adapt it for the *general* administration of medicines. It supplies a medium for administering active *fluid* medicines in a solid form while virtually in solution, having a range of utility as unlimited as that of the pilular form; it possesses the same advantages of concentration and portability, is more active because more quickly absorbed, is more easily dispensed, and is by far the most *agreeable* form of medicine which has hitherto been devised.

Another advantage which may fairly be claimed for the Glycecol over the ordinary hard lozenge, when employed in diseases of the mouth, tonsils, and throat, is its *soft and jelly-like* character, a valuable quality when the mucous membrane is inflamed. Being also free from sugar it is not liable to disorder the stomach, or induce nausea and loss of appetite, which is the usual effect of the common lozenge when frequently repeated,—and in throat affections frequent applications *are necessary.*

As a medium for administering medicine to children, Glycecols possess peculiar advantages which cannot be claimed for any other form; in most cases the taste of the medicine may be completely disguised, and even where nauseous substances are unavoidable, their objectionable flavour is materially lessened. They are attractive in appearance, infinitely more pleasant than any other vehicle, and are easily given to children too young to take the same medicines in the pilular form, and who invariably rebel against powders, which of all forms of medicine are the most nauseous. A Glycecol is a handsome and attractive looking lozenge, weighing from forty to sixty grains, lenticular in form, and about the size of a shilling, having the consistence of a firm jelly. When carrying a material soluble in Glycerine—Morphia for example—it is brilliantly transparent; in other cases, such as Guaiacum, it is opalescent; Bismuth and other insoluble substances, opaque. It will be observed that the active ingredients are more diluted than in the common lozenge, hence their complete diffusion and rapid absorption.

The following formulæ include the officinal troches and also those used in the *Hospital for Diseases of the Throat.* They may be much

F

extended, but the examples given are quite sufficient to indicate how almost universal is the application of this new vehicle.

MODE OF ADMINISTRATION.

WHEN EMPLOYED TO PRODUCE A LOCAL EFFECT the Glycecol should be placed upon the tongue and allowed to dissolve slowly; it can then be diffused over the part affected in the form of a liquid jelly. The time required for a Glycecol to dissolve is from four to five minutes, and as it is desirable to keep the medicament in contact with the part affected as long as possible this process should not be hurried. In severe cases of sore throat it is advantageous in order to remove any mucus or obnoxious secretion, that the mouth and throat should be first rinsed out with tepid water, to which, in most cases, a few drops of Condy's Fluid may be added.

WHEN USED AS A VEHICLE FOR MEDICINES designed to affect the system generally, the Glycecol as soon as it becomes softened, may be swallowed whole, before any flavour but that of glycerine is apparent.

WHEN ADMINISTERED TO CHILDREN. Glycecols may be given as above, or if the child be too young to be instructed, as in the case of infants, it should be placed in a silver teaspoon and held for a few minutes over a cup of hot water, and when dissolved should be given in a liquid state while yet *warm*, but on no account *hot*.

Glycecol Aconiti.*

℞ Tinct. Aconiti, B.P., ♏xij.; Glycerinæ, f℥ij.; Glycecolloidæ, ℨvj. M. ft. glycecol. xij.

Useful in painful affections of the pharynx; in phthisis they diminish expectoration and lessen the frequency of the pulse. Act as a sedative in irritative coughs. Dose, one every two hours. N.B.— Frequently repeated they produce constitutional effects.

---

* Dr. RINGER ("Handbook of Therapeutics," page 311) says,—"Of all the drugs we possess there are certainly none more valuable than Aconite. Its virtues are only beginning to be appreciated. . . . Its power over inflammation is little less than marvellous." In the early stages of acute inflammation, in pericorditis, pleurisy, pneumonia, tonsillitis, catarrh, rheumatism, and to relieve pain in neuralgia, nervous palpitation of the heart, etc., the preparations chosen will answer every purpose. For exhibiting minute doses, the Glycecol, Granules, or Tincture, are to be preferred.

Dr. RINGER prefers small doses often repeated; thus half a drop or a drop in a teaspoonful of water every ten minutes or quarter of an hour for two hours, and afterwards every hour. The Glycecols are admirably adapted for this purpose.

Glycecol Acidi Benzoici.

℞ Acidi Benzoici, gr. xij. vel vj. ; Glycerinæ, f ʒij. ; Glycecol-loidæ, ʒvj. M. ft. glycecol. xij.

A most valuable stimulant and voice lozenge, in cases of nervo-muscular weakness of the throat. Recommended to public speakers and professional singers.

Glycecol Acidi Carbolici.

℞ Acidi Carbolici Crys., gr. xij.; Glycerinæ, f ʒij.; Glycecol-loidæ, ʒvj. M. ft. glycecol. xij.

Useful in sore throat attended with fœtor of the breath.

Internally it is administered as an *antiseptic.* Its action is allied to that of Creosote, and it is given in similar cases.

Glycecol Acidi Gallici.

℞ Acidi Gallici, ʒj. ; Glycerinæ, f ʒij. ; Glycecolloidæ, ʒvj. M. ft. glycecol. xij.

Glycecol Acidi Tannici.

℞ Acidi Tannici, gr. xij. ; Glycerinæ, f ʒij. ; Glycecolloidæ, ʒvj. M. ft. glycecol. xij.

Useful for smokers.

Glycecol Acidi Tannici et Capsici.

℞ Acidi Tannici, gr. xij. ; Tr. Capsici, ♏xxxvj. ; Glycerinæ, f ʒij. ; Glycecolloidæ, ʒvj. M. ft. glycecol. xij.

Useful in relaxed throat and as a voice lozenge.

Glycecol Aloes.

℞ Ext. Aloes Pur., gr. vj., vel gr. iij. ; Ext. Glycyrrhizæ, ʒj. ; Glycerinæ, f ʒj. ; Glycecolloidæ, q. s. ft. ʒviij. M. ft. glycecol. xij.

A gentle laxative and remedy for habitual constipation. The taste of the Aloes is completely covered.

Glycecol Althææ.

℞ Pulv. Althææ, gr. xxxvj. ; Glycerinæ, f ʒij. ; Glycecolloidæ, ʒvj. M. ft. glycecol. xij.

This preparation is flavoured with orange flowers, and is in fact an improved form of *Pâte de Guimauve* with the addition of Glycerine. A harmless and demulcent pâte, well suited for delicate invalids, by whom it may be taken without fear of deranging the digestive organs.

Glycecol Aluminis.

℞ Aluminis, ʒj.; Glycerinæ, f ʒij.; Glycecolloidæ, ʒvj. M. ft. glycecol. xij.

More efficient than the Alum Gargle.

Glycecol Ammonii Bromidi.

℞ Ammonii Bromidi, gr. xxiv.; Glycerinæ, f ʒij.; Glycecolloidæ, vj. M. ft. glycecol. xij.

A useful remedy for hooping-cough.

Glycecol Ammonii Chloridi.

℞ Ammonii Chloridi, gr. xxxvj.; Glycerinæ, f ʒij. ; Glycecolloidæ, vj. M. ft. glycecol. xij.

Useful in bronchitis.

Glycecol Amygdalæ Amaræ.

℞ Pulv. Amygdalæ Amaræ Co., ʒss. ; Glycerinæ, f ʒj.; Glycecolloidæ, ʒiv. M. ft. glycecol. xij.

A mild sedative.

Glycecol Belladonnæ.*

℞ Succi Belladonnæ, f ʒj. ; Glycerinæ, f ʒj. ; Glycecolloidæ, ʒvj. M. ft. glycecol. xij.

Five minims of Belladonna juice is about equivalent to ⅛ gr. of Extract. Belladonnæ.

---

* This is a remedy of great value. Powerfully sedative, anodyne, and antispasmodic, it exercises a marked remedial influence in a large number of diseases,—as an antispasmodic in the advanced stages of hooping-cough (F. 31, 36 and 93), spasmodic asthma (F. 89, 33. Simple Glycecols or Granules),

Glycecol Bismuthi et Ammoniæ Citratis.

℞ Ammon. et Bismuthi Cit., gr. xxiv. ; Glycerinæ, f ʒj. ; Glycecolloidæ, ʒvj. M. ft. glycecol. xij.

Glycecol Bismuthi Subnit.

℞ Bismuthi Subnit., ʒj. ; Glycerinæ, f ʒjss. ; Glycecolloidæ, ʒvj. M. ft. glycecol. xij.

Glycecol Boracis.

℞ Boracis, ʒj. ; Glycerinæ, f ʒij. ; Glycecolloidæ, ʒv. M. ft. glycecol. xij.

Useful in thrush and aphthous ulcerations, fissured tongue, etc.

Glycecol Camphoræ.

℞ Camphoræ, gr. xij. ; Glycerinæ, f ʒij. ; Glycecolloidæ, ʒvj.; M. ft. glycecol. xij.

Glycecol Catechu.

℞ Ext. Catechu, gr. xxiv. ; Glycerinæ, f ʒij. ; Glycecolloidæ, ʒvj. M. ft. glycecol. xij.

Especially useful in sore throat and laryngeal irritation.

Glycecol Carbonis.

℞ Carbonis Ligni, ʒj. ; Glycerinæ, f ʒij. ; Glycecolloidæ, ʒv. M. ft. glycecol. xij.

---

laryngismus stridulus, chorea, epilepsy (F. 31), spasmodic stricture, chordee, dysmenorrhœa (F. 89); and as a sedative in cancer (F. 21). It is especially useful in incontinence of urine in children. (Simple Glycecols or Granules or F. 31.) As a remedial agent in scarlet fever it is of undoubted value, both as a curative and as a prophylactic measure.

It is well to bear in mind that Belladonna increases the pulse in force, frequency, and fulness (J. Harley); it is therefore a stimulant to the arterial system. Children bear large doses better than adults. The action of Belladonna should in any case be carefully watched, and its use be discontinued on the appearance of any of the following symptoms,—dilatation of the pupils, dryness of the mouth and fauces, thirst, and vertigo.

Glycecol Cerii Oxalatis.

℞ Cerii Oxalatis, ʒj.; Glycerinæ, f ʒij.; Glycecolloidæ, ad ʒj. M. ft. glycecol. xij.

Useful in irritable dyspepsia, attended with gastrodynia, pyrosis, and vomiting, especially the vomiting of pregnancy. It is also used in chorea, epilepsy, and other allied convulsive affections. Often found to succeed in cases where Nitrate of Silver and Bismuth have failed.

Glycecol Codeiæ.

℞ Codeiæ, gr. iij.; Glycerinæ, f ʒij.; Glycecolloidæ, ʒvj. M. ft. glycecol. xij.

Glycecol Colchici.

℞ Succi Colchici, f ʒj.; Glycecolloidæ, ʒv. vel ʒvj. M. ft. glycecol. xij.

Glycecol Conii.

℞ Succi Conii, f ʒj. vel f ʒij.; Glycerinæ f ʒj.; Glycecolloidæ, ʒvj. M. ft. glycecol. xij.

Glycecol Cubebæ.

℞ P. Cubebæ, gr. vj.; Glycerinæ, f ʒij.; Glycecolloidæ, ʒvj. M. ft. glycecol. xij.

Very serviceable in diminishing excessive secretions of mucus from the pharynx, larynx, or trachea. A modification of Brown's Bronchial Troches.

Glycecol Digitalis.

℞ Succi Digitalis, f ʒj.; Glycerinæ, f ʒj.; Glycecolloidæ, ʒvj. M. ft. glycecol. xij.

Glycecol Doveri.

℞ Pulv. Doveri, ʒss. vel ʒj.; Glycerinæ, f ʒij.; Glycecolloidæ, ʒv. vel ʒvj. M. ft. glycecol. xij.

Glycecol Doveri Mitius.

℞ Pulv. Doveri, gr. iij.; Glycerinæ, f 3ij.; Glycecolloidæ, 3vj. M. ft. glycecol. xij.

." A mild opiate for children from two to six weeks old."—DR. TANNER.

A soothing dose during painful dentition.

Glycecol Eucalypti.

℞ Ext. Eucalypti, gr. xij; Glycerinæ, f 3ij.; Glycecolloidæ, 3vj. M. ft. glycecol. xij.

Glycecol Ferri Redacti.

℞ Ferri Redacti, gr. xij. vel gr. xxiv.; Glycerinæ, f 3ij. Glycecolloidæ, 3vj. M. ft. glycecol. xij.

Glycecol Ferri Sacchar.

℞ Ferri Carb. Sacchar., gr. xxxvj.; Glycerine, f 3ij.; Glycecolloidæ, 3vj. M. ft. glycecol. xij.

Glycecol Guaiaci.

℞ Resinæ Guaiaci, gr. xxiv.; Glycerinæ, f 3ij.; Glycecolloidæ, 3vj. M. ft. glycecol. xij.

A specific for arresting crescent inflammation of the tonsils, and useful both in acute and subacute inflammations of the pharynx, and in acute follicular disease of the tonsils, etc.

Glycecol Gummi Rubri.

℞ Gummi Rubri, gr. xxiv.; Glycerinæ, f 3ij.; Glycecolloidæ, 3vj. M. ft. glycecol. xij.

A species of Kino. A very useful astringent in sore throat.

Glycecol Hæmatoxyli.

℞ Ext. Hæmatoxyli, 3j.; Glycerinæ, f 3j.; Glycecolloidæ, 3vj. M. ft. glycecol. xij.

Glycecol Hydrarg. c. Cretâ.

℞ Hydrarg. c. Cretâ, gr. xxxvj.; Glycerinæ, f 3ij.; Glycecolloidæ, 3vj. M. ft. glycecol. xij.

Glycecol Hyoscyami.

℞ Succi Hyoscyami, fȝj.; Glycerinæ, fȝj.; Glycecolloidæ, f ȝvj. M. ft. glycecol. xij.

Glycecol Ipecacuanhæ..

℞ Pulv. Ipecac., gr. iij.; Glycerinæ, f ȝij.; Glycecolloidæ, ȝvj. M. ft. glycecol. xij.

Glycecol Jacobi.

℞ Pulv. Jacobi, gr. iij. ; Glycerinæ, fȝij. ; Glycecolloidæ, ȝvj. M. ft. glycecol. xij.

Glycecol Krameriæ.

℞ Ext. Krameriæ. gr. xxxvj. ; Glycerinæ, fȝij. ; Glycecolloidæ, ȝvj. M. ft. glycecol. xij.

A very useful astringent. Rhatany does not disagree with the stomach as is often the case with Tannic Acid, nor does it cause constipation to the same extent as Kino or Catechu.

It is also serviceable in passive hæmorrhage, menorrhagia when not profuse but constant, in scorbutic affections of the mouth and gums, and in atonic dyspepsia.

Glycecol Lactucæ. T. H.

℞ Ext. Lactucæ, gr. vj.; Glycerinæ fȝij.; Glycecolloidæ, ȝvj. M. ft. glycecol. xij.

Glycecol Lobeliæ.

℞ Tr. Lobeliæ, fȝj.; Glycerinæ, fȝj.; Glycecolloid, ȝvj. M. ft. glycecol. xij.

Very useful in hooping-cough. The tincture is recommended by Dr. Ringer.

Glycecol Lithiæ.

℞ Lithiæ Citratis, ȝj.; Glycerinæ, fȝij.; Glycecolloidæ, ȝv. M. ft. glycecol. xij.

Given with much benefit in gout, rheumatism, and other affections occurring in connection with the Lithic Acid diathesis. It has an alkaline influence on the urine, and is a solvent of Lithic Acid deposits, hence its value in cases where Urate of Soda is liable to be deposited in the tissues.

Glycecol Lupulinæ.

℞ Lupulinæ, ʒj.; Glycerinæ, f ʒj.; Glycecolloidæ, ʒvj. M. ft. glycecol. xij.

A mild tonic and sedative; very useful in atonic dyspepsia.

Glycecol Morphiæ. B. P.

℞ Morphiæ Mur., gr. j.; Glycerinæ, f ʒvj.; Glycecolloidæ, ʒxviij. M. ft. glycecol. xxxvj.

Glycecol Nucis Vomicæ.

℞ Ext. Nucis Vomic., ½ gr.; Glycerinæ, f ʒij.; Glycecolloidæ, ʒvj. M. ft. glycecol. xij.

Glycecol Opii.

℞ Ext. Opii, gr. j.; Glycerinæ, f ʒjss.; Glycecolloidæ, ʒv. M. ft. glycecol. x.

Glycecol Pepsinæ.

℞ Pepsinæ Porci, gr. xxiv.; Glycerinæ, f ʒij.; Glycecolloidæ, ʒvj. M. ft. glycecol. xij.

The best and most effective mode of administering pepsin.

Glycecol Piperis Nigri.

℞ Tinct. Piperis Comp. =Conf. Pip. Nig., ʒij.; Glycerinæ, ʒij.; Glycecolloidæ, ʒiv. M. ft. glycecol. xij.

A modification of Ward's Paste. A useful remedy in hæmorrhoids.

Glycecol Podophylli.

℞ Resinæ Podophylli, gr. j.; Glycerinæ, f ʒj.; Glycecolloidæ, ʒiij. M. ft. glycecol. 50.

Prepared especially for infants and young children. Relieves obstinate constipation.

Notwithstanding the "uncertainty" which repute attributes to the action of this drug, it is daily growing into favour. It is constantly prescribed in all those cases in which Calomel or some other mercurial was formerly employed. It is exceedingly useful in liver affections and disordered secretions, in skin affections (especially eczema), and whenever an alterative is required. Dr. Ringer prescribes it in obstinate constipation, which sometimes occurs during the early months of infant life, with hard motions of a clay colour. In such cases he directs one or two drops of a tincture to be given twice or thrice a day. Podophyllin Glycecols are prepared for this same purpose. These may be given with great safety at any period of infant life. The formulæ numbered 10, 53, 60, 119, 128 and 129 are all active and efficient preparations.

Glycecol Potassæ Chloratis.

℞ Potassæ Chloratis, ʒj.; Pulv. Althææ xxxvj.; Glycerinæ, ʒj.; Glycecolloidæ, ʒvj. M. ft. glycecol. xij.

Antiseptic. Useful locally in thrush and aphthous ulcerations of the mouth and fauces, and as a general tonic to the mucous membrane. Often repeated produce constitutional effects. Useful in typhus and scarlet fevers.

Glycecol Potassæ Nitratis.

℞ Potassæ Nitratis, gr. xxxvj.; Glycerinæ, fʒij.; Glycecolloidæ, ʒvj. M. ft. glycecol. xij.

Glycecol Pruni Virg.

℞ Pulv. Cort. Pruni Virg. gr. xxiv.; Glycerinæ, fʒij.; Glycecolloidæ, ʒvj. M. ft. glycecol. xij.

The glycecol developes hydrocyanic acid when introduced into the mouth, acting as a sedative.

Glycecol Pulvis Cretæ Aromaticæ.

R Pulv. Cretæ Aromat., ʒj.; Glycerinæ, fʒj.; Glycecolloidæ, ʒvj. M. ft. glycecol. xij.

Glycecol Pulvis Kino Comp. = ⅛ gr. Opium.

℞ Pulv. Kino. Co., ʒss.; Glycerinæ, fʒjss.; Glycecolloidæ, ʒvj. M. ft. glycecol. xij.

Glycecol Pulvis Opii Comp. = ¼ gr. Opium.

℞ Pulv. Opii Co., ʒss.; Glycerinæ, fʒjss.; Glycecolloidæ, ʒvj. M. ft. glycecol. xij.

A very valuable and convenient remedy for diarrhœa and summer cholera.

Glycecol Quiniæ.

℞ Quiniæ Sulph., gr. iij.; Glycerinæ, f℥ij.; Glycecolloidæ, ℥vj. M. ft. glycecol. xij.

Glycecol Rosæ.

A harmless and delicious pâte, well adapted for invalid children. In throat affections and fevers it is a pleasant and nutritious form of demulcent.

Glycecol Santonini.

℞ Santonini, xxiv.; Glycerinæ, f℥ij.; Glycecolloidæ, ℥vj. M. ft. glycecol. xij.

Especially useful in the treatment of *lumbricus* or round worm. The combination with Scammony, F. 43, is a very useful one, and seldom fails. It is also useful in the treatment of *ascarides*, the cure of which is greatly facilitated by the exhibition of Sulphate of Iron and Quassia, F. 44, twice or thrice daily.

Glycecol Scammonii.

℞ Scammonii Virg., gr. xxxvj.; Glycerinæ, f℥ij.; Glyce-colloidæ, ℥vj. M. ft. glycecol. xij.

Glycecol Senegæ.

℞ Tr. Senegæ Conc., ℥ij.; Glycerinæ, f℥j.; Glycecolloidæ, ℥vj. M. ft. glycecol. xij.

A valuable stimulating expectorant.

Glycecol Sennæ Co. = Confection of Senna one drachm in each.

℞ Conc. Comp. Tinct. Sennæ = Conf. Sennæ Co., ℥jss; Glycerinæ, f℥j.; Glycecolloidæ, ℥v. M. ft. glycecol. xij.

A mild and efficient aperient, useful during gestation.

Glycecol Sodæ Carb.

℞ Sodæ Carb. Exsic., ℥j.; Glycerinæ, f℥ij.; Glycecolloidæ, ℥vj. M. ft. glycecol. xij.

Glycecol Sodæ Sulphitis.

℞ Sodæ Sulphitis, gr. xxiv.; Glycerinæ, f℥ij.: Glycecolloidæ, ℥vj. M. ft. glycecol. xij.

Used to eliminate Sulphurous Acid; destructive to vegetable life; a specific in aphthæ.

Glycecol Zingiberis.

℞ Tr. Zingib. Fort., f℥j.; Glycerinæ, f℥j.; Glycecolloidæ, ℥vj. M. ft. glycecol. xij.

For the composition, &c., of the following, see FORMULÆ, PILULÆ. They are especially adapted, with few exceptions, to the treatment of diseases of infancy and childhood.

Glycecol Aloes et Jalapinæ. *F.* 66.

Glycecol Aloes et Nucis Vomicæ. *F.* 67.

Glycecol Aloes et Pepsinæ. *F.* 45.

Glycecol Antim. Tart. et Doveri. *F.* 98.

Glycecol Belladonnæ et Ammon. Bromidi.

℞ Succi Belladonnæ, ♏xxxvj.; Ammon. Bromidi, gr. xij.; Glycerinæ, f℥ij.; Glycecolloidæ, ℥vj. ·M. ft. glycecol. xij.

A very excellent remedy for hooping cough.

Glycecol Belladonnæ et Zinci Sulph. *F.* 31.

This combination, in the form of granules, is a favourite remedy for hooping-cough with many practitioners. In this form it has many advantages.

Glycecol Belladonnæ et Potassii Bromidi. *F.* 93.

Glycecol Bismuthi et Hyoscyami. *F.* 49.

Glycecol Calomel et Antim. *F.* 112.

Glycecol Camphoræ Co. *F.* 79.

Glycecol Camph. Co. et Ipecac. *F.* 123.

A very useful anodyne expectorant medicine, each glycecol being equivalent to ♏xv. Tr. Camph. Co. This is a very useful cough medicine and an excellent substitute for Paregoric.

Glycecol Camphoræ et Potassæ Nit.

℞ Camphoræ, gr. xij.; Potassæ Nitratis, gr. xxxvj.; Glycerinæ, f ℨij.; Glycecolloidæ, ℨvj. M. ft. glycecol. xij.

Glycecol Doveri et Hyd. c. Cretâ. *F.* 92 *and F.* 149.

Glycecol Hydrarg. c. Cretâ Doveri, et Sodæ. *F.* 39.

Glycecol Hydrarg. c. Cretâ, Rhei, et Ipecac. *F.* 27.

Glycecol Ipecac., Morphiæ, et Scillæ. *F.* 96.

A very favourite remedy in the pilule form.

Glycecol Ipecac. et Morphiæ.

℞ P. Ipecac. gr. iij.; Morphiæ Mur., gr. j.; Glycerinæ, f ℨvj.; Glycecolloidæ, ℨxviij. M. ft. glycecol. xxxvj.

Equivalent to the B.P. Troche.

An elegant preparation, very useful in bronchitic and phthisical coughs.

Glycecol Ipecac., Aconiti, et Potassæ Chloratis. *F.* 103.

Glycecol Ipecac., Potassæ Nit., et Papaveris. *F.* 102.

Glycecol Santonini et Scammonii. *F.* 43.

Glycecol Scammonii et Calomel. *F.* 42.

Glycecol Scammonii, Calomel, et Jalapinæ. *F.* 41.

Glycecol Sodæ Carb. et Zingib. _F._ 109.

Glycecol Sodæ Carb., Rhei, et Zingib.

℞ Sodæ Carb. Exsic., gr. xxiv.; Ext. Rhei, gr. vj.; P. Zingib., gr. vj.; Glycerinæ, f ʒij.; Glycecolloidæ, ʒvj. M. ft. glycecol. xij.

Glycecol Sodæ Carb., Rhei, et Hyd. c. Cretâ. _F._ 188.

———

POTUS.
Refrigerant and Saline Drinks.

In the treatment of many acute diseases, and especially in fevers of the eruptive class, it is desirable to exhibit Salines and Refrigerant Medicines in a fluid form. The Mineral Acids, and Salts of Potash, Magnesia, and Soda, are all useful, and are constantly prescribed. When it is not convenient to administer them in the form of a mixture, I order them in the household form of _a drink_, which I find patients, especially little ones, take more willingly than "regular physic"; there is no difficulty in getting them quickly and well prepared. Medicine in this form, supplemented by Glycecols, Granules, or Pills, efficiently supplies the requirements of most cases. It is to be remembered that the best and most effective refrigerant is cold or iced water.

Potus Acidi Hydrochlorici.

℞ Acidi Hydrochlorici Dil., f ʒij. vel f ʒiij.; Mellis Depurati, ʒj.; Decocti Hordei, Oij. Mix for the daily drink.

"In typhus fever."—DR. TANNER.

The diluted Nitro-hydrochloric and Phosphoric Acids may in some cases replace the Hydrochloric Acid.

Potus Potassæ Bicarbonatis.

℞ Potassæ Bicarb., ʒij.; P. Sacchari Alb., ʒij.; Ol. Limonis, ♏iij. M. ft. pulv.

To be dissolved in one or two pints of water according to the condition of the patient, and taken as a drink in 24 or 36 hours. 'n acute rheumatism.

Potus Potassæ Chloratis.

℞ Potassæ Chloratis, ʒij.; P. Sacchari Alb., ʒij. M. ft. pulv.

To be dissolved in two pints of barley water or lemonade, and taken in the course of 24 hours, half a tumblerful for a dose. Very useful in scarlet and typhoid fevers.

Potus Salina.

℞ Sodii Chlor., gr. xx.; Potassæ Chloratis, gr. vij.; Sodæ Carb., gr. xxx. M. ft. pulv.

To be taken every half hour largely diluted. Used by Dr. Stevens in the saline treatment of malignant cholera.

Potus Chlori vel Mistura Chlorinii.
(*Middlesex.*)

℞ Solutio Chlorini,* ʒij.; Syrupi, ʒj.; A'quæ. ad ʒviij.

This constitutes the Mistura Chlorinii of the Middlesex Hospital, the dose of which is one to two tablespoonfuls. Useful in typhus and scarlet fevers; also as a gargle in malignant sore throat.

Potus Potassæ Tartratis Acida (Imperial Drink).
(*St. Bartholomew's.*)

℞ Potassæ Bitart., ʒij.; Ol. Limonis, m v.; Aquæ Bullientis, ad ʒxx.

Nutritious Demulcent Drinks.

Mix together half a pint of Mucilago Acaciæ, Mistura Amygdalæ, and pure milk; sweeten with sugar-candy or honey; and add one large tablespoonful of any liqueur. Allow the whole to be taken during the day. Or a large pinch of isinglass may be boiled with a tumblerful of milk, half a dozen bruised almonds, and two or three lumps of sugar. To be taken warm once or twice in the day.

These drinks are very grateful in cases of tonsillitis, ulceration of the pharynx, etc., also in some cases of debility with irritability of the stomach and a tendency to diarrhœa.

---

* **Solutio Chlori** (*Middlesex*).

℞ Chlorate Potash, 30 grs.; Hydrochloric Acid, ʒss.; Water, ad ʒj.

PART II.—EXTERNAL AND TOPICAL APPLICATIONS.

This is a very important class of remedies, comprising a great variety of applications of undoubted value. Ointments and plaisters have fallen much into disuse of late years. They have been replaced by many elegant and useful applications. Simple water dressing, or water medicated with Permanganate of Potash, (Condy's Fluid), Fluid Opium, the Chloride or Sulphate of Zinc, are all excellent dressings for granulating wounds, and give no trouble in their preparation.

COLLYRIA.

Collyrium Ammon. Acet.

℞ Liq. Ammon. Acetatis, ʒj. ; Aquæ Dest., ad ʒvij.

Collyrium Aluminis.

℞ Alum. gr. ij. to iv. ; Water, ʒj.

Collyrium Argenti Nit.

℞ Nitrate Silver, gr. ij.; Water, ʒj.

Collyrium Argenti Nit. Fort.*

℞ Nitrate Silver, gr. iv.; Water ʒj. *London Ophthalmic.*

Collyrium Atropiæ.

℞ Sulphate Atropia, gr. ij. ; Water, ʒj.

---

* In the Pharmacopœia of the London Ophthalmic Hospital these formulæ are classed under the head of Guttæ. I have placed them under that of Collyria, which term is more indicative of their use; they are, however, to be *dropped* into the eye—not used as eye *washes*.

Collyrium Atropiæ Fort.*

℞ Sulphate Atropia, gr. iv. ; Water, ℨj.

Collyrium Cupri Sulph.

℞ Sulphate Copper, gr. j. ; Water, ℨj.

Collyrium Opii Co.

℞ Tr. Opii, ℨj. ; Sol. Ammon. Acet., ℨss. ; Distilled Water, ad ℨiv. *Guy's:* Guttæ.

Collyrium Plumbi Acet.

℞ Acetate Lead, gr. ij. ; Dilute Acetic Acid, ♏j. ; Water, ℨj. *Middlesex, Westminster.*

Collyrium Potassii Iodidi.

℞ Potassii Iodidi, gr. iij.; Water, ℨj. *London Ophthalmic.*

Collyrium Zinci Acet.

℞ Acetate Zinc, gr. ijss.; Water, ℨj.

Collyrium Zinci et Aluminis.

℞ Sulphate Zinc, gr. ijss. ; Alumen, gr. ij. ; Water, ℨj.

Collyrium Zinci Chloridi.

℞ Chloride Zinc, gr. j. ; Water, ℨj.

Collyrium Zinci Sulph. c. Opio.

℞ Sulphate Zinc, gr. ij. ; Tr. Opii, ♏ xx. ; Water, ℨj.

Collyrium Zinci Sulph.

℞ Sulphate Zinc, gr. j. ; Water, ℨj.

---

* Same strength as the Liquor Atropiæ Sulph., B.P. Solutions of Atropia should be freshly prepared for use, as they spoil by keeping.

G

LINIMENTA.

The Liniments of the Pharmacopœia are all very useful and excellent remedies, leaving nothing to be desired in that form of application. Those in use at the hospitals are inferior, the formulæ being evidently framed with a view to economy.

Linimentum Camphoræ.

℞ Olive Oil, f ℥iv.; Camphor, ℥j. Dissolve.

This is a useful Stock Liniment. It affords a ready means of quickly preparing the following :—

Linimentum Chloroformi.

℞ Chloroformi, Linimenti Camphoræ, equal parts.

Linimentum Crinale.

℞ Ol. Amygdalæ Dulc., f ℥j. ; Liq. Ammon. Fort., f ℥j. ; Aquæ Mellis, f ℥ij. ; Sp. Rosmarini, f ℥iv. Mft. Linimentum.

A useful stimulating liniment for the scalp. May be employed with advantage in baldness and to prevent the falling off of the hair after fevers.

Linimentum Hydrargyri.

℞ Ung. Hydrargyri, Liquor Ammoniæ, Linimentum Camphoræ, aa f ℥j.

Linimentum Terebinthinæ Aceticum.

℞ Sp. Terebinthinæ, Linimenti Camphoræ, Acidi Acetici, equal parts.

Linimentum Ammoniæ.

℞ Liquor Ammoniæ, 1 part; Linimenti Camphoræ, 3 parts.

Linimentum Crotonis.

℞ Ol. Crotonis, ℥j.; Linimenti Camphoræ, ℥j.

Linimentum Opii.

℞ Tinct. Opii, Linimenti Camphoræ, equal parts.

Linimentum Iodi will be found amongst the Pigmenta; Linimenta Aconiti, Belladonnæ, Potassii Iodidi c. Sapone, Saponis, and Sinapis Co., require special preparation, and should therefore be obtained from a house of repute ready made.

Equal *parts of Olive Oil and Lime Water* forms the Linimentum Calcis of the British Pharmacopœia.

PIGMENTA.

SOLUTIONS FOR LOCAL APPLICATION.

THROAT PIGMENTS to be used with a brush are to be preferred to gargles in affections situated behind the anterior pillars of the fauces. Suitable camel's hair brushes are made for these and similar applications. *

From the Pharmacopœia of the Hospital for Diseases of the Throat.

Pigmentum Acidi Carbolici.

30 grains of the crystals in each fluid ounce of water.
Antiseptic.

Pigmentum Aluminii Chloridi.

15 minims of the solution in each fluid ounce of water.
Antiseptic and astringent.

† Pigmentum Argenti Nitratis.

60 grains of the salt in each fluid ounce of water.

Pigmentum Cupri· Sulphatis.

15 grains of the salt in each fluid ounce of water.

Pigmentum Ferri Aluminis.

60 grains of the salt in each fluid ounce of water.
Astringent.

Pigmentum Ferri Perchloridi Fort.

120 grains of the salt in each fluid ounce of water.
· Astringent, hæmostatic.

---

* For weaker solutions used for *Atomised Inhalations*, see page 106.
† Solutions of nitrate of silver are not recommended by Dr. Morell Mackenzie for general topical application to the larynx except in cases of tertiary syphilitic ulceration. . . . Even in the case of specific ulcers of the throat the solution of sulphate of copper is as efficacious as that of nitrate of silver, and far less frequently creates spasm or nausea.

Pigmentum Ferri Perchloridi Dilut.

60 grains of the salt in each fluid ounce of water.

Astringent.

Pigmentum Ferri Sulphatis.

60 grains of the salt in each fluid ounce of water.

Astringent.

Pigmentum Zinci Chloridi Fort.

30 grains of the salt in each fluid ounce of water.

Astringent.

Pigmentum Zinci Sulphatis.

60 grains of the salt in each fluid ounce of water.

PIGMENTA, OR PAINTS, FOR EXTERNAL APPLICATION.

Pigmentum Guttæ Perchæ.

(Fever Hospital.)

℞ Guttæ Perchæ, ʒjss.; Chloroformi, ʒj.

Used to prevent pitting in small-pox and to paint over superficial excoriations, threatened bedsores, etc., etc.

Pigmentum Collodii c. Iodo.

(Middlesex Hospital.)

℞ Iodi, gr. xij.; Potasii Iodidi, gr. xxiv.; Collodii, f ʒj.

Pigmentum Collodii Tincti Preparati.

℞ Collodii, ʒj.; Ol. Palmæ, gr. x.; Rad. Anchusæ, q.s.

Pigmentum Iodi.

(London Hospital.)

℞ Iodi, gr. xl.; Potassii Iodidi, gr. xx.; Sp. Vini. Rect., ʒjss.

With the addition of Camphor, gr. x., this closely resembles the *Linimentum Iodi, B.P.*

Pigmentum Iodi.

(King's College Hospital.)

℞ Iodi. ʒj.; Sp. Vini Rect., ℥jss.

Pigmentum Olei Ricini c. Collodio.

(Fever Hospital.)

℞ Collodii, ʒiv.; Ol. Ricini, ℥j.

Used for superficial burns. This is an excellent remedy. Should be applied with a broad varnish brush.

Pigmentum Ovi.

(Fever Hospital.)

℞ Albumen Ovi, no. ij.; Sp. Vini Rect., ℥j.

For cure and prevention of bed sores.

Pigmentum Sulphuris.

(King's College Hospital.)

℞ Sulph. Precip., Pot. Carb., Glycerinæ, Sp. Vini Rect., āā ℥j.

A specific for itch.

GLYCERINA.

Glycerinum Acidi Carbolici Mitius.

(King's College Hospital.)

℞ Acidi Carbolici, ʒj.; Glycerinæ, ad ℥j.

A capital remedy for ringworm.

Glycerinum Acidi Gallici.

(St. Bartholomew's Hospital.)

℞ Acidi Gallici, gr. xl.; Sp. Vini Rect., ʒij.; Glycerinæ, ad f ℥j.

Glycerinum Acidi Tannici.

(London Hospital.)

℞ Acidi Tannici, ʒiij.; Glycerinæ, ʒvj.; Sp. Vini Rect., Aquæ Dest., āā ʒvj.

Glycerinum Belladonnæ.

(St. Bartholomew's Hospital.)

℞ Ext. Belladonnæ, ʒj.; Glycerinæ, ʒiv.

Glycerinum Ferri Perchloridi.

(King's College Hospital.)

℞ Sol. Ferri Perchlor., ʒss.; Glycerinæ, ʒss.

A saturated solution in glycerine is made by H. & T. Kirby & Co. It is a powerful styptic, and otherwise a useful pigment.

Glycerinum Zinci.

A combination of Carbonate and Oxide of Zinc in Glycerine, a valuable pigment in the treatment of vesicular affections of the skin, an almost specific in eczema. It was largely prescribed by the late Mr. Startin.

UNGUENTA.

Unguentum Simplex.

℞ White Wax, ʒij.; Prepared Lard, ʒiij.; Almond Oil, f ʒiij.

Melt the wax and lard in the oil on a water-bath, then remove the mixture, and stir constantly until it cools.

This Ointment or Prepared Lard should be kept in stock, either of which may be used as a base for the following Ointments. They are readily prepared by triturating the active ingredients in a porcelain mortar, and gradually adding the ointment or lard, rubbing them well together until perfectly smooth and free from grittiness.

Unguentum Aconitiæ.

℞ Aconitiæ, grs. viij.; Unguenti Simplicis, ʒj.

Unguentum Sulphuris Iodidi.

℞ Sulph. Iodidi, ʒss.; Unguenti Simplicis, ʒj.

Unguentum Creosoti.

℞ Creosoti, f ʒj.; Unguenti Simplicis, ʒj.

Unguentum Cadmii Iodidi.

℞ Cadmii Iodidi, 62 grs.; Unguenti Simplicis, ℥j.

Unguentum Hydrarg. Ammon.

℞ Hydrarg. Ammon., 62 grs.; Unguenti Simplicis, ℥j.

Unguentum Plumbi Carb.

℞ Plumbi Carb., 62 grs.; Unguenti Simplicis, ℥j.

Unguentum Plumbi Iodidi.

℞ Plumbi Iodidi, 62 grs.; Unguenti Simplicis, ʒj.

Unguentum Hydrargyri Oxidi Rub.

℞ Hyd. Oxid. Rub., 62 grs. ; Yellow Wax ʒij. ; Almond Oil, ʒvj.

Melt the wax at a gentle heat, mix the oil with it, and when nearly cold add the mercury, and mix thoroughly.

Unguentum Belladonnæ.

℞ Ext. Belladonnæ, 80 grs.; Unguenti Simplicis, ℥j.

Unguentum Hydrargyri Subchloridi.

℞ Calomel, 80 grs.; Unguenti Simplicis, ℥j.

Unguentum Antimonii Tartarati.

℞ Antimonii Tartarati, ¼ oz.; Unguenti Simplicis, ℥j.

Unguentum Gallæ.

℞ Pulv. Gallæ, 80 grs.; Unguenti Simplicis, ℥j.

Unguentum Zinci Oxidi.

℞ Zinci Oxidi, 80 grs.; Unguenti Simplicis, ℥j.

Unguenta Hydrargyri Nit., Hydrargyri, Iodi, Resinæ, Sabinæ, and others not mentioned, require special preparation, and should be obtained of some house of repute, care being taken that they be ordered *freshly prepared.*

CAUSTICA.

Causticum Argent. Nit.

> ℞ Nitrate Silver, 1 dram; Spirits Nitric Ether, 1 oz.; Indigo, 5 grs. *Skin.*

Used for phlegmonous, vesiculous, and ulcerous affections.

Causticum Arseniosum Co.

> ℞ Arsenious Acid, 1 dram; Vermilion, 2 scruples; Calomel, 2½ grs. *Skin.*

Used in ulcerous, strumous, and cachectic affections.

> ℞ Arsenious Acid, Powdered Acacia, of each 1 oz.; Water 5 fluid drams.

Dr. Marsden speaks highly of this caustic in epithelioma. The affected part is to be painted over with it night and morning, taking care rigorously to limit the application to the diseased parts, and not to let it extend over more than one superficial inch at a time. As the part sloughs, its separation is to be aided by bread and water poultices; and when all the disease has been got rid of by the repeated applications of the mucilage, a carrot poultice is to be applied during the night, and a weak black wash (Calomel 60 grs. to Lime Water, 1 pint) during the day, until the part is healed.

Causticum Hydrargyri Perchloridi.

> ℞ Corrosive Sublimate, 1 dram; Prepared Coloured Collodion, 6 drams.

Useful in structural, tubercular, and ulcerous affections.

Causticum Sabinæ Co.

> ℞ Powdered Savin, 1 dram; Burnt Alum, 15 grs.; Levigated Red Precipitate, 15 grs. *Skin.*

Used in structural affections.

Causticum Zinci Chloridi.

> ℞ Chloride Zinc and Oxide Zinc, equal parts. *London Ophthalmic.*

> ℞ Powdered Bloodroot, ½ oz. to 1 oz.; Chloride Zinc, ½ oz. to 2 oz.; Water, 2 oz.; Farina, sufficient to make a paste. Mix.

The paste thus formed should have the consistence of treacle. This is the caustic which was employed by Dr. Fell.

Causticum Zinci Supersulph.

℞ ½ fluid oz. of Sulphuric Acid, and saturate it with Sulphate Zinc, previously dried and powdered.

Sir J. Y. Simpson recommends that this caustic should be used by dipping a pen in it, and then drawing lines across the tumour, so as to eat through the skin in a few minutes. The fissures thus made are to be filled with the paste, renewing the scratching and caustic every day or two. In this way five or eight days may suffice for the removal of a good-sized tumour. By this combination also we can penetrate deeply, without hardening the parts and without fear of producing hæmorrhage. This is a very valuable caustic, and the author has found it particularly useful for the removal of cancerous tumours of the breast, etc. The pain which it produces will be best mitigated by employing the subcutaneous injection of morphia at each application. See *Index of Diseases*, Dr. Tanner, page 332.

London Paste.

(*Throat Hospital.*)

℞ Caustic Soda and Unslaked Lime of each equal parts.

Reduce to a fine powder in a warm mortar, and mix intimately. Keep in well closed bottles, and when required for use take as much as is sufficient, and make into a paste with water.

Recommended for destroying enlarged tonsils or the elongated uvula, where treatment with the guillotine or scissors is objected to.

This preparation resembles the Vienna Paste, but is preferable in consequence of its being less liable to spread beyond the limits of application. Soda being used instead of potash, and water in place of alcohol, the preparation is much less painful.

INJECTIONS, URETHRAL.

Injectio Acidi Tannici.

℞ Acid. Tannic., gr. ij.; Water, ℥j.

Injectio Opii.

℞ Liq. Opii Sed., ♏ x.; Water, ℥j.

Injectio Zinci Chlor.

℞　Chloride Zinc, gr. j.; Liq. Opii Sed. ℳ x.; Water, ℥j.

Injectio Zinci Sulph.

℞　Sulphate Zinc, gr. j.; Liq. Opii Sed. ℳ x.; Water, ℥j.

Mr. Durham has devised a syringe for urethral injections which will at once commend itself to the judgment of the reader. The annexed drawing completely illustrates its action.

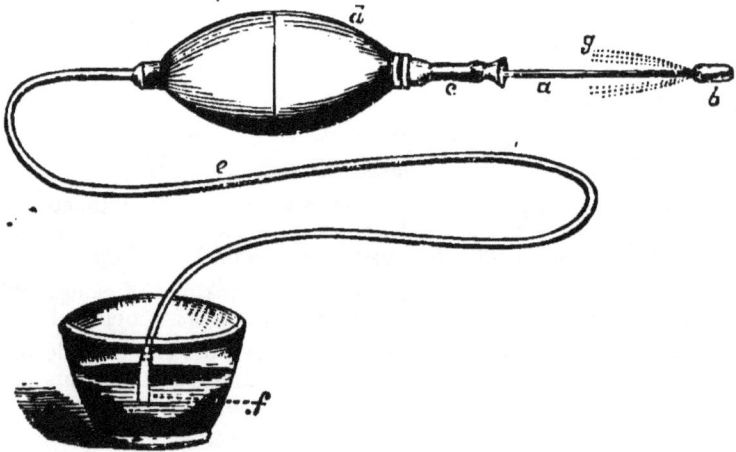

The extremity (*f*) of the tube (*e*) is sunk in the fluid to be injected, and the syringe is filled. The tube (*a*) is then introduced, previously oiled, into the Urethra, until the bulb has passed beyond the point to which the inflammation extends. The hand-ball is then compressed, and the fluid issues through perforations made in such direction that it flows from backwards, as indicated in the figure.

Thus a gentle stream *from within outwards* is maintained, and infectious matter is prevented from being carried on towards the bladder by the bulb of the instrument, which is grasped by the

Urethra. The mucous membrane can be washed free from dis-
charge, and astringent or soothing injections continuously applied.

INJECTIONS, VAGINAL.

Injectio Vaginalis Astringens.

℞ Tannic Acid, ʒj.; Alum, ʒij.; Water, Oj. *University.*

This is an effective and convenient substitute for the well-known
Alum and Oak Bark Injection, commonly used in the hospitals.

Injectio Vaginalis Astringens.

℞ Alum, ʒss.; Sulphate Zinc, ʒss.; ℞ ft. pulv.

To be dissolved in Oj. water, for one injection, and applied by
means of Higginson's, or the Pneumatic Syringe, as shown below.

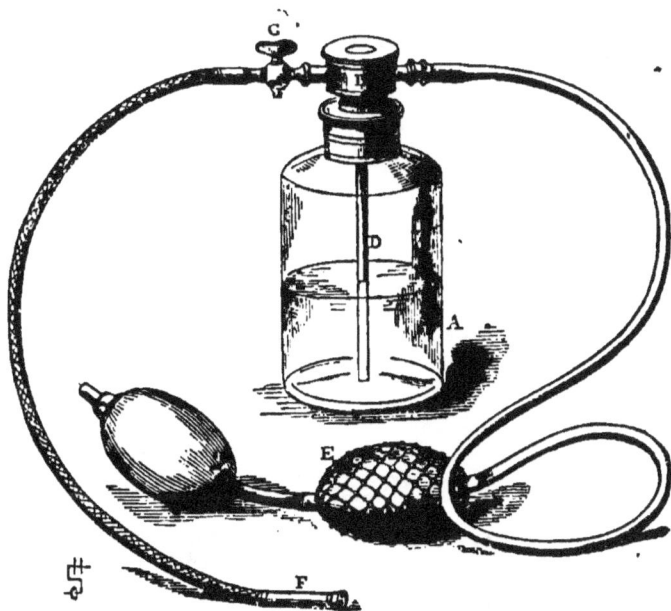

By this contrivance a *continuous* stream is maintained.

BOUGIA.

MEDICATED BOUGIES (Soluble). Diameter similar to a No. 9 Catheter. Length about 2 inches.

Successfully employed in the treatment of gleet, and in gonorrhœa after the first inflammatory symptoms have subsided.

Bougia Acidi Gallici.

Gallic Acid, 1 gr.

Bougia Acidi Tannici.

Tannic Acid, 1 gr.

Bougia Argenti Nitratis.

Nitrate of Silver, $\frac{1}{2}$ gr. and $\frac{1}{4}$ gr.

Bougia Belladonnæ.

Alcoholic Extract of Belladonna, $\frac{1}{2}$ gr.

Bougia Cupri Sulph.

Sulphate of Copper, 1 gr.

Bougia Ferri Perchloridi.

Perchloride of Iron, 1 gr.

Bougia Opii.

Opium, 2 grs.

Bougia Zinci Sulphatis.

Dried Sulphate of Zinc, 1 gr.

PESSARIA.

Vaginal Pessaries are very useful and portable remedies, and may often be employed in the place of injections with advantage.

Pessaria Acidi Carbolici.

Carbolic Acid (deodorant), gr. v.

Pessaria Acidi Gallici.
Gallic Acid (astringent), gr. x.

Pessaria Acidi Tannici.
Tannic Acid, gr. x. (astringent). *London.*

Pessaria Aluminis.
Alum (astringent), gr. xv.

Pessaria Aluminis et Zinci.
Dried Alum, gr. v. ; Sulphate of Zinc, gr. v. ; Opium, gr. ij.; Oil of Theobroma, ʒ j. *London.*

Pessaria Belladonnæ.
Alcoholic Ext. Belladonna (sedative), gr. ij.

Pessaria Ferri Perchloridi.
Perchloride of Iron (hæmostatic), gr. v.

Pessaria Hydrargyri.
Mercurial (alterative and resolvent) (*Ung. Hydrarg.*), gr. xxx.

Pessaria Iodoform.
Iodoform, gr. x. ; Oil of Theobroma, gr. lx.

Pessaria Morphiæ.
Hydrochlorate of Morphia, gr. $\frac{1}{2}$; Oil of Theobroma, gr. lxx. *London.*

Pessaria Opii.
Opium (sedative), gr. ij.

Pessaria Plumbi Acetatis.
Acetate of Lead (astringent), gr. 7$\frac{1}{2}$.

Pessaria Plumbi et Opii.
Acetate of Lead, gr. v.; Opium, gr. ij (astringent).

Pessaria Plumbi Iodidi et Atropiæ.

Iodide of Lead, gr. x.; Sulphate of Atropia, gr. ₁'₈; Oil or Theobroma, ℥ j. *London.*

Pessaria Potassii Bromidi.

Bromide of Potassium (alterative and resolvent), gr. x.

Pessaria Zinci et Atropiæ.

Dried Sulphate of Zinc, gr. x.; Sulphate of Atropia, gr. ₁'₈; Oil of Theobroma, ℥ j. *London.*

Pessaria Zinci Oxidi.

Zinc, Oxide (cicatrising and emollient), gr. xv.

Pessaria Zinci Sulphatis.

Sulphate of Zinc (dried) caustic, gr. x.

SUPPOSITORIA.

Suppositorium Acidi Carbolici.

Acid Carbolic, gr. j.; Curd Soap, gr. xv.; Starch, a sufficiency. *B.P.*, 1874.

Suppositorium Acidi Tannici.

Tannic Acid (*B.P.*, 1867), gr. iij.; Cocoa Butter, gr. xv.

Acid Tannic, gr. iij.; Curd Soap, gr. viij½.; Glycerine of Starch, gr. iv¼.; Starch, a sufficiency. *B.P.*, 1874.

Suppositorium Belladonnæ.

Belladonna, gr. ij. (Ext.); Cocoa Butter, gr. xv.

Suppositorium Elaterii.

Extract of Elaterium, gr. ij.; Soap, gr. x.; Flour, gr. x.; water, q.s. *St. Bartholomew's.*

Suppositorium Gallæ et Opii.

Galls, gr. v.; Opium, gr. j.; Cocoa Butter, gr. xv.

Suppositorium Hydrargyri.

Mercurial Ointment, gr. v.; Cocoa Butter, gr. xv. (*B.P.*, 1867).

Suppositorium Hyoscyami.

Extract Henbane, gr. v.; Cocoa Butter, gr. xv.

Suppositorium Morphiæ.

Morphia, gr. ½ and gr. ¼; Cocoa Butter, gr. xv. (*B.P.*, 1867).
Muriate of Morphia, gr. ½; Curd Soap, gr. viijss.; Glycerine of Starch, gr. iv⅛.; Starch, a sufficiency. *B.P.*, 1874.

Suppositorium Opii.

Opium, gr. ij.; Cocoa Butter, gr. xv.

Suppositorium Plumbi et Opii.

Acetate Lead, gr. iij.; Opium, gr. j.; Cocoa Butter, gr. xv. (*B.P.*, 1867).

Suppositorium Podophylli.

Podophyllin, gr. j.; Cocoa Butter, gr. xv.

Suppositorium Santonini.

Santonine, gr. v.; Cocoa Butter, gr. xv.

Suppositorium Crotonis.

Croton Oil, 5 mins.; Crumb of Bread, gr. xxx. *Westminster.*

Suppositorium Zinci Oxidi.

Zinc Oxide, gr. x.; Cocoa Butter, gr. xv.

Suppositorium Zinci Sulphatis.

Zinc Sulphate (dried), gr. iij.; Cocoa Butter, gr. xv.

FOTUS.

Fotus Ammoniæ Acet. c. Opio.

℞ Liq. Ammoniæ Acet. Conc., f℥vj.; Ext. Opii. Liquidi, f℥ij.
vel Tr. Opii., f℥iv.

To be put into half a pint of hot water, and used as a fomentation.

This application the writer has found singularly useful. It is a
powerful discutient, and in orchitis it is especially serviceable. For
severe *injuries*, sprains and bruises, in local inflammations, gout or
rheumatism it affords considerable relief.

Fotus Belladonnæ.

℞ Extract of Belladonna, gr. lx.; distilled water, oz. xx.
London Ophthalmic. St. Bartholomew's.

Fotus Belladonnæ c. Opio.

℞ Ext. Belladonnæ, Ext. Opii., aa. gr. 90 ; Glycerinæ, f℥iv.;
Ext. Papaveris, ℥jss. Mix.

To be painted over the seat of inflammation in pleurisy, peritonitis, gastric disease, etc. A fomentation flannel, a hot linseed
poultice, or wet compress is to be applied ; being separated from the
extracts by a sheet of tissue paper.—DR. TANNER.

Fotus Terebinthinæ.

Hot damp flannel, sprinkled with Oil of Turpentine. *Fever.*

CATAPLASMATA.

Cataplasma Acidi Carbolici.

Make a Linseed Poultice, but substitute the Carbolic Acid
Lotion * for one half of the water. *Fever.*

Cataplasma Carotæ.

Carrots boiled until they are soft, and scraped into a pulp.
Westminster.

---

* One part of Carbolic Acid (Crystals) to 55 of water.

Cataplasma Iodi.

Linseed Poultice sprinkled with Tincture of Iodine. *Fever.*

Cataplasma Panis.

Grated bread and boiling water, of each sufficient. *London.*

Cataplasma Plumbi.

Solution of Subacetate of Lead 1; water 1; bread, q.s. *Westminster.*

Cataplasma Plumbi et Opii.

Linseed Poultice substituting Lead and Opium Lotion for half the water. *Fever.*

Cataplasma Sinapis.

Linseed Cataplasm sprinkled with Mustard. *Guy's.*

BALNEA.

Balneum Acidum.

Nitric Acid, 1½ fl. oz.; Hydrochloric Acid, 1 fl. oz.; water 30 gals. *Skin. University. St. Bartholomew's.*

Used in chromatic, papular, and squamous affections.

Balneum Acidum Nitro Hydrochloricum.

Nitric Acid, 11 fl. oz.; Hydrochloric Acid, 20 fl. oz.; boiling water, 30 gals. *Guy's.*

Nitric Acid, 15 fl. oz.; Hydrochloric Acid, 30 fl. oz.; water, 30 gals. *London.*

Balneum Alkalinum.

Crystal. Carb. Soda, 4 oz.; hot water, 30 gals. *Skin.*

Crystal. Carb. Soda, 6 oz.; water, 30 gals. *London University.*

Carbonate of Soda Crystals, 8 oz.; water, 30 gals. *St. Bartholomew's.*

Carbonate of Potash, 2 oz.; warm water, 25 gals. *Middlesex.*

Used in phlegmonous, papular, squamous, and corneous affections.

H

Balneum Boracis Co.

Borax, 2 oz. ; Precipitated Sulphur, 2 oz.; hot water, 30 gals. *Skin.*

Used in parasitic and pustular affections.

Balneum Conii Co.

Ext. Conium, 2 oz. ; Starch, 1 lb. ; boiling water, 1 gal. ; boil a short time, and add water to 30 gals. *Skin.*

Used in phlegmonous, papular, and neuralgic affections.

Balneum Creasoti.

Creasote, ¼ oz. ; Glycerin, 2 oz. ; hot water, 30 gals. *Skin. St. Bartholomew's.*

Used in vesicular, papular, and squamous affections.

Balneum Glycerinæ Co.

Glycerin, 2 oz.; Tragacanth, 2 oz. ; boil in a pint of water, and add water to 30 gals. *Skin.*

Used in vesicular, corneous, and squamous affections.

Balneum Iodi.

Iodine, ¼ oz.; Solution of Potash, 2 oz. ; water, 30 gals. *Skin.*

Used in tubercular, cachectic, and squamous affections.

Balneum Marinum vel Sodii Chloridi.

Bay Salt, 8 lbs. ; water, 30 gals. *London.*

Bay Salt, 9 lbs.; water, 30 gals. *St. Bartholomew's.*

Bay Salt, 8 oz.; Sulphate of Magnesia, 2 oz.; Solution of Chloride of Calcium, 1 oz.; water, 30 gals. *Skin.*

Used in chromatic, papular, and sebaceous affections.

Balneum Mercuriale.

Corrosive Sublimate, 3 drms.; Hydrochloric Acid, 1 drm.; water, 30 gals. *Skin.*

Used in tubercular, cachectic, and squamous affections.

Balneum Potassæ Sulphuratæ vel Sulphureum.

Sulphurated Potash, 4 oz.; boiling water, 30 gals. *London* and *Guy's.*

Sulphurated Potash, 2 oz.; Hyposulphite of Soda, 1 oz.; Dilute Sulphuric Acid, ½ oz.; water, 30 gals. *University.*

Sulphurated Potash, 8 oz.; warm water, 25 gals. *Middlesex. St. Bartholomew's.*

Balneum Sulphuris Co.

Precipitated Sulphur, 2 oz.; Hyposulphite of Soda, 1 oz.; diluted Sulphuric Acid, ½ oz.; water, 30 gals. *Skin.*

Used in vesicular, papular, squamous, and parasitic affections.

Balneum Calidum Aeris Madefacti,—

		110° to 115° F.
,,	Calidum	98° to 110° F.
,,	Tepidum	85° to 92° F.
,,	Frigidum	56° to 64° F.
		Guy's. London.
,,	Calidum	96°.
,,	Tepidum	90°.
		St. Bartholomew's.

Mercurial Vapour Baths.

"The patient is seated on a chair, and covered with an oil-cloth lined with flannel, which is supported by a proper frame-work. Under the chair are placed a copper bath containing water, and a metallic plate on which is put from sixty to one hundred and eighty grains of the bisulphuret of mercury, or the same quantity of the grey oxide or the red oxide of this metal. In syphilitic affections of the skin, testes, and bones, from five to thirty grains of the green iodide of mercury may be employed; or a mixture of twenty grains of the green iodide with ninety grains of the bisulphuret often proves efficacious. Under the bath and plate spirit lamps are lighted. The patient is thus exposed to the influence of three agents—heated air, steam, and the vapour of mercury. At the end of five to ten minutes perspiration commences, which becomes excessive in ten or fifteen minutes longer. The lamps are now to be extinguished, and when the patient has become moderately cool he is to be

rubbed dry. He should then drink a cup of warm decoction of
guaiacum or sarsaparilla, and repose for a short time.—LANGSTON
PARKER. *In constitutional syphilis, when mercury is indicated.
This method of introducing mercury into the system may also be
adopted with benefit in other diseases, in place of administering the
metal by the mouth.*

The Turkish Bath.

The general effect of a hot air bath is to increase the force and
rapidity of the circulation, and to induce free perspiration; but
if too hot or too prolonged, the determination of blood to the
skin and lungs becomes so great that the brain suffers. There is
then, consequently, a lowering of the circulation, with depressed
nervous power. A temperature varying from 120° to 165° will
usually suffice; while if the perspiration is efficient and continuous,
and the sensation agreeable, the patient may remain in the cali-
darium for from forty to sixty minutes. The bath is useful in
removing local congestions, in clearing the pores, and in inducing a
healthy condition of the skin and mucous membranes, in elimi-
nating noxious matters from the blood, and in imparting a sense of
elasticity and vigour to the system. It is *injurious* when there is
any obstruction to the circulation, or when the heart or vessels are
affected with fatty degeneration, or when there are symptoms of
disease of the nervous centres, or when there is a tendency to vertigo
or syncope, and in advanced life. Women who are pregnant, or
who are menstruating, ought not to have recourse to it.

Cold Affusion.

The patient is seated in an empty bath, and from four to six
buckets of cold water (about 40° F.) are poured over his head and
chest from a height of two or more feet. He is then quickly dried,
and replaced in bed. The colder the water and the greater the
height from which it is poured, the more stimulating the effect..
Affusion as thus practised by Dr. Currie proved very valuable in
the treatment of typhus. It may be resorted to when the temperature of the body is permanently above its normal (about 98° F.)
standard, when there is no feeling of chilliness, when the body is
not wholly bathed in sweat, when there is not much irritability of the
nervous system, and when there is great stupor. The effect is to
lower the temperature, to lessen the frequency of the pulse and
respiration, to render the tongue moist and soft, to diminish or remove
stupor, to procure sleep, and sometimes to produce a critical perspiration. It may be used every twenty-four hours if necessary.

The Douche Bath.

When it is desirable to apply this form of bath to one or more of the joints, it is only necessary to affix two or three yards of large sized india-rubber tubing to the tap of a cistern. The patient must sit in an empty bath into which the water may fall as it plays upon the limb.

Wet Sheet Packing.

The patient is closely enveloped in a sheet which has been dipped in cold or tepid water and well wrung out. He is then carefully wrapped in a blanket, covered with three or four more blankets, and a down coverlet is tucked over all. He should remain thus for thirty, forty-five, or sixty minutes lying on his side, or in a semi-recumbent position; the duration being timed by the sedative effect produced. The sweating is not generally excessive; but the water, urea, and chloride of sodium of the urine are slightly increased, this increase being considerable when the sheet is continued for four hours.

The Blanket Bath.

This affords an easy means of inducing sweating. A blanket is wrung out of hot water and wrapped round the patient. He is to be packed in three or four dry blankets, and allowed to repose for thirty minutes. The surface of the body should then be well rubbed with warm towels, and the patient made comfortable in bed.

The Wet Compress.

This consists merely of a roll of flannel or calico, dipped in cold water and wrung out, and then applied around the seat of pain. Over this a piece of waterproof cloth is to be worn.

The Warm Bath as a Cooling Agent.

The warm bath at a temperature of 95° F. must prove a cooling agent to the body of a fever patient at 100° to 105° F. The immersion should continue from fifteen minutes to an hour or longer. Its sedative effects render it valuable where the nervous system is irritable.

In cases of delirium tremens with high fever, *cold superfusion* may be used while the patient is held in the warm bath. From ten to thirty buckets of cold water are to be poured slowly over the head; hot water being continually added to the bath to maintain its heat at 95° F. This treatment generally produces sound sleep.

DR. TANNER.

Acid Sponging.

One part of vinegar is to be added to two or three of cold water, and the body well sponged with the mixture. Simple tepid water may sometimes be advantageously used. The patient being weak and unable to move, the sponging must be done by degrees, *i.e.*, the arms, chest, back, and legs are to be rapidly washed and dried. Useful in many cases of fever, inflammation, scarlatina, etc.

———

ENEMATA.

The *dissolving* power of the rectal fluids is very inferior to that of the stomach, and absorption also takes place more slowly. Thus medicines intended to produce a constitutional effect should be in solution. They require a longer time to affect the system than when given by the mouth. Food requires special preparation.

Enema Assafœtidæ.

Tincture of Assafœtida, ʒiv.; Decoction of Barley, oz. xx. *Fever. London.*

Enema Commune.

Chloride of Sodium, oz. j.; Decoction of Barley (*Guy's* Decoction of Oats), oz. xij. *Westminster. Guy's.*

Decoction of Barley, oz. xx. *Middlesex.*

Chloride of Sodium, oz. j.; Decoction of Barley, oz. xx. *St. Thomas's. London.*

Enema Olei Ricini.

Castor Oil, oz. ij.; Decoction of Barley (tepid), oz. viij. *St. Thomas's. Charing Cross.*

Castor Oil, oz. ij.; Starch, ʒj.; Decoction of Oats, oz. xij. *Westminster.*

Castor Oil, oz. j.; Honey, oz. j.; Decoction of Oats (tepid), oz. x. *Guy's.*

Castor Oil, oz. ij.; Mucilage of Starch, oz. xviij. *London. St. Bartholomew's.*

Enema Olei Ricini c. Assafœtidâ.

Castor Oil, ℥ iv. ; Tincture of Assafœtida, ʒiv. ; Mucilage of Starch, to oz. xij. *Fever.*

Enema Oleosum.

Olive Oil, oz. iv. ; Decoction of Oats, oz. xij. *St. George's.*
Olive Oil, oz. iv. ; Decoction of Barley, oz. xvi. *Middlesex.*
Olive Oil, oz. iv. ; Mucilage of Starch, oz. xvi. *London.*

Enema Opii.

Tincture of Opium, 30 drops ; Mixture of Starch, oz. ij.—
Consumption. Tincture of Opium, 15 mins. ; Mucilage of Starch, oz. ij.—*Fever.*

Enema Spiritus Vini Gallici.

Brandy, oz. j. ; Strong Beef Tea, oz. iij. *Fever.*

Enema Tabaci, B.P.

Leaf Tobacco, 20 grains; Boiling Water, 8 ozs. Infuse half an hour and strain.
Inject oz. viij., and, if necessary, repeat it in an hour. *Guy's.*

Enema Terebinthinæ.

Oil of Turpentine, oz. j. ; Mucilage of Starch, oz. xx. *Fever.*

NUTRITIVE ENEMATA.
Beef Tea and Brandy.

Take six ounces of strong beef tea, an ounce of cream, and half an ounce of brandy or an ounce and a half of port wine. This may be administered twice or thrice in the course of twenty-four hours. In cases of acute gastritis, carcinoma of the stomach, obstinate vomiting, etc., where it is necessary to avoid giving food by the mouth.

Opium, Iron, and Quinine.

Take four or six ounces of restorative soup (F. 2), one ounce of cream, two teaspoonfuls of brandy, ten or fifteen minims of liquid extract of opium, and ten grains of citrate of iron and quinia.

Cod Liver Oil and Bark.

Take four ounces of essence of beef (F. 3), two ounces of port wine, an ounce of cod liver oil, two drachms of tincture of yellow cinchona, and twenty minims of liquid extract of opium. Mix. To be administered every twelve hours.

Quinine and Beef.

Take one tablespoonful of brandy, five grains of sulphate of quinine, one teaspoonful of glycerine, two tablespoonfuls of cream, and four to eight ounces of restorative soup, F. 2. Mix. This enema may be administered every six or eight hours. Where the rectum is very irritable, or it is necessary to relieve pain, from fifteen to twenty minims of the liquid extract of opium may be advantageously added.

Feeding by the Rectum.—It is sometimes of the greatest consequence to feed a patient in other ways than by the stomach; and therefore the question of nourishing by the rectum is one of the gravest importance. The solvent influence of the rectal juices is practically nothing, and it is necessary to digest the food before injecting it into the rectum. This may be done in the following manner:—The pancreas of a bullock is to be finely chopped, freed from fat, and mixed with eight or nine ounces of glycerine; a third part of this mixture, when about to be used, is added to five ounces of finely chopped meat and should be injected into the rectum as soon as it is made.

VAPORES.—INHALATIONS.

(From the Pharmacopœia of the Hospital for Diseases of the Throat.)

Inhalations as here prescribed are of five kinds:—

1. STEAM INHALATIONS; *i.e.*, steam impregnated with volatile matter. Temperature, 130 F. to 150° F.

2. COLD INHALATIONS. Temperature, 60° F. to 100° F.

3. DRY INHALATIONS; *i.e.*, volatile matters vaporized by heat.

4. ATOMISED INHALATIONS; *i.e.*, inhalations of atomised fluids.

5. FUMING INHALATIONS; *i.e.*, inhalations of the smoke of ignited nitrated papers.

(1.) Steam Inhalations.

The value of steam inhalations has long been recognised both by the profession and the public. The curative effect of this class of remedies is, no doubt, in part due to the steam, but a special character is imparted to them by the particular medicament employed in addition to the hot water.

In the subjoined formulæ, the quantities of ingredients are generally prescribed for three ounce mixtures, a teaspoonful of which is added to a pint of water at the required temperature, for each inhalation. Although formulæ are given for each medicament, the quantity of the volatile oil may be increased according to the circumstances of the case, and it is often desirable to combine several oils or other remedies in the same prescription. In the case of most of the essential oils, light carbonate of magnesia is used to hold the oil in suspension, in the proportion of half a grain of magnesia to each drop of the oil. This medium is preferable to mucilage, glycerine, or spirit of wine.

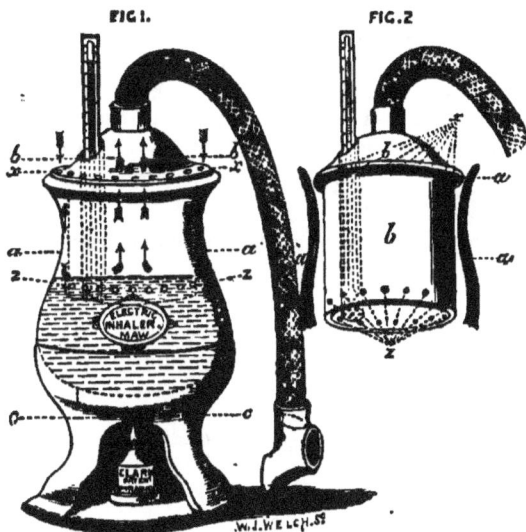

ELECTIC INHALER. Recommended by Dr. Morrel Mackenzie.

(2.) Cold Inhalations.

Cold inhalations are indicated when it is desirable to produce a general effect on the mucous membrane of the throat and where hot inhalations cause headache and faintness. The temperature may vary from 60° to 100° F. Cold inhalations are also useful in hot seasons and hot climates. Any of the forms recommended for cold inhalation can, if it be desired, be employed at a high temperature, but in that case it is generally necessary to slightly reduce their strength. For cold inhalations, the Eclectic Inhaler answers equally well as for steam inhalations.

(3.) Dry Inhalations.

Dry hot inhalations are indicated in cases of excessive secretion, but are difficult of administration, as it is almost impossible to raise the temperature, in any small inhaler, to a sufficient degree, without a very complicated apparatus. By a slight adaptation of Messrs. Bullock & Reynold's Eclectic Inhaler, however, that apparatus may be conveniently employed.

(4.) Atomised Inhalations.

"Atomised Medicated Fluids may be advantageously used in affections of the lining membrane of the nose, mouth, and fauces, in Croup and Diphtheria, Syphilitic affections of palate and throat, Laryngitis, Œdema of the Glottis, Hooping Cough, Bronchitis, Phthisis, Hoarseness, and Loss of Voice. During the application, the parties should make deep and long inspirations and expirations. Except in acute cases one application daily will suffice."—TANNER'S INDEX OF DISEASES, p. 341.

The following are the principal remedies used as Atomised Inhalations :—

```
 *  Aqua Acidi Carbolici.  .   .   .  30 grs. to 10 ozs. water.
 †    ,,     ,,   Lactici  .   .   .  f ʒijss.      ,,       ,,
      ,,     ,,   Sulphurosi  .  .   50 min.       ,,       ,,
      ,,     ,,   Tannici  .  .  50 to 200 grs.    ,,       ,,
      ,,   Calcis, B.P.
      ,,   Aluminii Chloridi  .  f ʒj. solution    ,,       ,,
      ,,   Aluminis .  .  .  .  .   80 grs.        ,,       ,,
      ,,   Ferri Aluminis .  .  · .  30 ,,         ,,       ,,
      ,,     ,,   Perchloridi  .  .  30 ,,         ,,       ,,
      ,,     ,,   Sulphatis  .  .  20 to 40 ,,     ,,       ,,
      ,,   Potassæ Permanganatis  .  50 ,,         ,,       ,,
      ,,   Sodii Chloridi  .  .  .  50 ,,          ,,       ,,
      ,,   Zinci Chloridi  .  .  20 to 50 ,,       ,,       ,,
      ,,     ,,   Sulphatis .  .  .  50 ,,         ,,       ,,
```

The inhalation of atomised fluids is acknowledged to be a rational and successful method of treating various affections of the nose, throat, air passages, and lungs. By its means the indicated remedy is applied directly to the diseased tissues, and is, in these cases, what washes and lotions are to the exposed surfaces when diseased or mechanically injured. Inhalation may in this manner

---

* Especially valuable where there is a deficient secretion of mucus.
† This remedy has been found of great service in diphtheria; it appears to have the effect of dissolving the membranous exudation.

supplement constitutional treatment by the ordinary means, and it will be found a valuable auxiliary. Biegel's Practical Treatise "On Inhalation as a means of Local Treatment, etc.," will supply the reader with very complete information on the subject.

Siegle's apparatus acts exceedingly well, the spray is warm, and so fine that it causes little or no irritation on the most sensitive surface; it is self-acting and neither fatigues the patient nor requires an assistant. Moreover, it is handy, durable, portable, and its cheapness brings it within the reach of all classes.

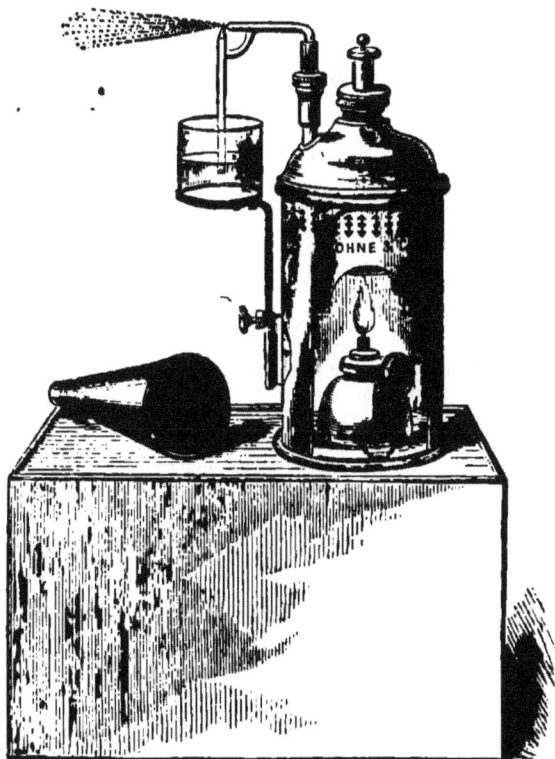

(5.) Fuming Inhalations.

These inhalations are derived from smoke arising from the ignition of unsized paper steeped in a solution of nitrate of potash. Though the value of fuming inhalations has long been recognised both by physicians and patients, the remedy has not hitherto been placed on a scientific basis. This is now done by requiring the papers to be steeped in solutions of definite strength, and by modifying their effects by the addition of various volatile principles.

This form of inhalation is recommended in cases of spasmodic dyspnœa, especially when dependent on asthmatic complications or on spasm of the adductors of the vocal cords. The method of using the papers is as follows :—a strip is lit at one end and dropped into a cylindrical vessel about four inches high and of a diameter of two inches. A wire gauze cover is then put on, and the fumes are inhaled by repeated deep inspirations.

Vapor Acidi Acetici.

℞ Acetic Acid, Glacial Acetic Acid, ot each, f ʒjss.　Mix.

Two teaspoonfuls in a pint of water at 140° F. for each inhalation.

Sedative, antiseptic, and is very useful in the inflammatory sore throat of scarlet fever, etc.

Vapor Acidi Carbolici.

℞ Carbolic Acid, ʒxxj.; Water, f ʒiij.

For steam inhalation use as above. For cold inhalation, two teaspoonfuls in a pint of water at 80° F. to 100° F.

Antiseptic. Very serviceable in syphilitic and carcinomatous ulcerations.

Vapor Acidi Hydrocyanici.

℞ Acid Hydrocyanic Dilute, B.P., fʒiij.; Water, ad f ʒiij.

One teaspoonful in a pint of water at 80° F. for each inhalation.

Sedative. Very useful in the cough of laryngeal phthisis and in some spasmodic affections.

Vapor Acidi Sulphurosi.

℞ Sulphurous Acid, f ʒj.; Water, f ʒxx., for each inhalation.

The temperature of this inhalation may vary from 60° to 100° F.

Stimulant. The value of this remedy has been much overestimated both by the public and the profession. It is apt to cause spasmodic irritation of the air tubes. It should be inhaled VERY SLOWLY.

Vapor Ætheris.

℞ Ether, Rectified Spirit, of each, f ℥jss.

One teaspoonful in a pint of water at 80° F. for each inhalation.

Vapor Ætheris Acetici.

℞ Acetic Ether, Rectified Spirit, of each, f ℥jss.

One teaspoonful in a pint of water at 140° F. for each inhalation.

It may also be used as a cold inhalation at 80° F.

Sedative. Often useful in irritation of the larynx.

Vapor Aldehydi.

℞ Dilute Aldehyde, f ℥iv.; Water, ad ℥iij.

One teaspoonful in a pint of water at 150° F. for each inhalation.

Sedative. Useful in recent catarrhal congestions and as a nasal inhalation in ozæna. It is contra-indicated in cases of asthma.

Vapor Ammoniæ.

℞ Liquor Ammoniæ, B.P., sp. grav. ·959, f℥j. ad f ℥jss.; Water, f℥iij. Mix.

One teaspoonful of this solution in a pint of water at 80° F. for each inhalation.

Stimulant; useful in chronic laryngitis and functional aphonia.

This inhalation may be advantageously employed in combination with any of the volatile oils, Camphor, or Thymol.

Vapor Amyl Nitritis.

℞ Nitrite of Amyl, 24 minims; Rectified Spirit, ad f ℥iij.

One teaspoonful in a pint of water at 100° F. for each inhalation.

Antispasmodic. Very valuable in some cases of asthma and spasm of the glottis.

Vapor Benzoini.

℞ Compound Tincture of Benzoin.

One teaspoonful in a pint of water at 150° F. for each inhalation.

A most valuable sedative inhalation for acute inflammations of the pharynx and larynx, especially in their early stages.

Vapor Calami Aromatici.

℞ Oil of Sweet Flag, 16 minims; Light Carbonate Magnesia, 8 grs.; Water, ad f ℥iij.

One teaspoonful in a pint of water at 150° F. for each inhalation.

A powerful stimulant. It often acts admirably in cases of chronic congestion of the larynx, when other stimulating inhalations have lost their effect.

Vapor Camphoræ.

℞ Sp. Camphor, f ʒiij.; Rectified Spirit, f ʒj.; Water, ad ℥iij.

One teaspoonful in a pint of water at 150° F. for each inhalation. *To be inhaled slowly.*

Stimulant. Very valuable in cases of chronic glandular laryngitis.

Vapor Chloroformi.

℞ Chloroform, Rectified Spirit, of each, f ℥jss.

A teaspoonful to be added to a pint of water at the desired temperature (from 60° to 100° F.), and an additional teaspoonful to be added every five minutes during the time that the inhalation is used. Not more than three teaspoonfuls to be used on any single occasion except in the presence of a medical practioner.

Sedative. Gives great relief in hay fever and in spasmodic affections of the larynx.

Vapor Creosoti.

℞ Creosote, f ℥ss.; Light Carbonate Magnesia, 90 grs.; Water, ad f ℥iij.

One teaspoonful in a pint of water at 150° F. for each inhalation.

Stimulant. A very serviceable remedy for chronic congestion of the larynx and trachea. Also of great use in ozæna.

Vapor Iodi.

Pour ten drops of Tincture of Iodine into the apparatus for dry inhalation, and inhale the vapor; in most cases it is desirable to add a fresh quantity of the tincture twice or thrice on each occasion of inhaling.

Stimulant. Useful where pus is formed in large quantities. It sometimes restores the voice in functional aphonia. It is also recommended in some forms of hay fever.

Vapor Lupuli.

℞ Oil of Hops, 20 minims; Light Carbonate Magnesia, 30 grs.; Water, ad ʒiij.

One teaspoonful in a pint of water at 150° F. for each inhalation.

Sedative. Especially useful in relieving the distressing cough of laryngeal phthisis.

Note.—The above formulæ are selected from the Pharmacopœia of the Hospital for Diseases of the Throat.

Vapor Acidi Carbolici.

Crystals of Carbolic Acid, 30 grs.; boiling water, 20 oz. *Chest.*

Vapor Æther. Chlor. c. Hyoscyami.

Chloric Ether, 30 mins.; Tincture of Henbane, 30 mins.; hot Infusion of Hop or water, 8 oz. *Consumption.*

Vapor Camphoræ.

Spirits of Camphor, 1 to 2 drms.; boiling water, 8 oz. *Consumption.*

Vapor Chloroformi.

Chloroform, 15 mins. for one inhalation. *Consumption.*

Vapor Conii Succi.

Juice of Conium, ½ oz.; boiling water, 8 ozs. *Consumption.*

Vapor Creasoti.

Creasote, ⅓ drm.; Mucilage, ⅓ drm.; hot water, 10 oz. *London.*

Vapor Iodi.

Tincture of Iodine, 40 mins.; hot water, 10 oz. *London.*

Vapor Opii.

Extract of Opium, 3 grs.; hot water, sufficient. *Consumption.*

HYPODERMIC INJECTIONS.*

Injection of Aconitine.

℞ Aconitine, gr. j. ; Sp. Vini Rect., ♏x. ; Aquæ Dest., ad
f ʒij. Mix.

For the first injection not more than two minims should be
employed. The dose may afterwards be safely increased to four
minims ($\frac{1}{30}$ gr.). It is better, though not absolutely necessary, to
make the injection at the seat of pain.

Injection of Atropine.†

The subcutaneous injection of Atropine is sometimes useful in
cases of intestinal obstruction, asthma, tetanus, neuralgia, chorea in
the adult, etc. *Great caution* is necessary. Not more than two
minims of the officinal Liquor Atropia = $\frac{1}{30}$ gr. should be employed
at first.

Injection of Morphia, B.P., 1874.†

A solution of Acetate of Morphia, containing 1 grain of the
Acetate in 12 minims of the injection. For first injections not more
than three minims should be used, as it is certain that this narcotic
acts more powerfully when thus employed than when taken into the
stomach. In diseases which are continuously painful, the ease given
by an injection will last for about twelve hours. To relieve the
suffering of advanced cancer, etc., the injection may be advantageously given night and morning for many months.

Injection of Chloroform.

An injection of ten or fifteen minims often effects a cure for the
time in pleurodynia, neuralgia, sciatica, etc. It has the disadvantage of sometimes producing an irritable ulcer, which may be slow
in healing.—TANNER'S " Index of Diseases."

---

* It may for convenience be stated in this place that the primary dose of a salt
of Morphia should never exceed one eighth, of Atropine one sixtieth, of Strychnia
one twenty-fourth, of Aconitine one thirtieth, of a grain, and of Quinine one or
two grains.—Dr. STILLE.
 † Atropine and Morphia combined may be used in neuralgic pains about the eye.

Part III.—Aliments, Etc.

ALIMENTS.*

(1.) Extract of Beef.

Take one pound of rump steak, mince it like sausage meat, and mix it with one pint of cold water. Place it in a pot at the side of the fire, to heat very slowly. It may stand two or three hours before it is allowed to simmer, and then let it boil gently for fifteen minutes. Skim and serve. The addition of a small teaspoonful of cream to a teacupful of this beef renders it richer and more nourishing. Sometimes it is preferred when thickened with a little flour or arrowroot.

(2.) Restorative Soup for Invalids.

Take one pound of newly-killed beef or fowl, chop fine, add eight fluid ounces of soft or distilled water, four or six drops of pure hydrochloric acid, thirty to sixty grains of common salt, and stir well together. After three hours the whole is to be thrown on a conical hair sieve, and the fluid allowed to pass through with slight pressure. On the fresh residue in the sieve, pour slowly two ounces of distilled water, and let it run through while squeezing the meat. There is thus obtained about ten fluid ounces of cold juice (cold extract of flesh of red colour), possessing a pleasant taste of soup; of which a wineglassful may be taken at pleasure. It must not be warmed, at least not to a greater extent than can be effected by standing in hot water a bottle partially filled with the juice, since it is rendered muddy by heat or by alcohol, and deposits a thick coagulum of albumen with the colouring matter of blood. If from any special circumstances (such as a free secretion of gastric juice), it is deemed undesirable to administer an acid, the soup may be well prepared by merely soaking the mincemeat in plain distilled water. Children will frequently take the raw meat simply minced, when they are suffering from great debility. One tea-spoonful of such meat may be given every three hours.

---

* For many excellent receipts the reader is referred to the work of Miss Acton, Crefyeld's *Family Fare*, and Dr. Tanner's *Index of Diseases*, from which some of the formulæ have been taken.

I

This modification of Liebig's formula is very valuable in cases of continued fever, in dysentery, and indeed in all diseases attended with great prostration and weakness of the digestive organs. When the flavour is thought disagreeable, it may be concealed by the addition of spice, or of a wineglassful of claret to each teacup of soup.

(3.) Essence of Beef.*

Take one pound of gravy beef, free from skin and fat, chop it up as fine as mincemeat, and pound it in a mortar with two tablespoonfuls of soft water. Then put it into a covered earthen jar with a little salt, cementing the ridges of the cover with pudding or paste. Place the jar in an oven, or tie it tightly in a cloth, and plunge it into a pot of boiling water for three hours. Strain off (through a coarse sieve so as to allow the smaller particles of meat to pass) the liquid essence, which will amount to about two ounces in quantity. Give two or more teaspoonfuls frequently.

In great debility, diphtheria, exhaustion from hæmorrhage, etc.

(4.) Brandy and Egg Mixture.

Take the white and yolks of three eggs, and beat them up in four ounces of plain water, add slowly three or four ounces of brandy, with a little sugar and nutmeg. This form is preferable to that in "London Pharmacopœia;" for 1851.

Two tablespoonfuls should be given every four or six hours. In some cases of great prostration the efficacy of the mixture is much increased by the addition of one drachm of the tincture of yellow cinchona to each dose.

(5.) A Special Restorative.

Dr. Dobell.

New milk, 4 parts; beef tea, cold, 2 parts; pale brandy, 1 part.

If no other food is taken, about five ounces (half an ordinary tumblerful) should be given every two hours, or half that quantity every hour. When desirable this food may be gradually thickened by the addition of *boiled* corn flour or other farinaceous articles, and flavoured with spice.

(6.) Stewed Oysters.

Half a pint of oysters, half an ounce of butter, flour, one third of a pint of cream, cayenne and salt to taste. Scald the oysters in their own liquor, take them out, beard them, and strain the liquor. Put the butter into a stewpan, dredge in sufficient flour to dry it up,

---

* Dr. Leared has suggested an admirable apparatus for making *bee tea*. Messrs. Maw Son & Thompson are the makers.

add the oyster liquor and stir it over a sharp fire with a wooden spoon; when it comes to a boil add the cream, oysters, and seasoning; let all simmer for one or two minutes, but *not longer*, or the oysters will harden; serve on a hot dish, with croutons or toasted sippets of bread. A quarter of a pint of oysters, the other ingredients in proportion, make a dish large enough for one person.

(7.) Panada.

Take the crumb of a penny roll, and soak it in milk for half an hour, then squeeze the milk from it; have ready an equal quantity of chicken or veal scooped very fine with a knife; pound the bread crumbs and meat together in a mortar. It may be cooked either mixed with veal or chicken broth, or by taking it up in two teaspoons in pieces the shape of an egg after seasoning it, poached like an egg, and served on mashed potato.

(8.) Macaroni.

Two ounces of macaroni, a quarter of a pint of milk, a quarter of pint of good beef gravy, the yolk of one egg, two tablespoonfuls of cream, half an ounce of butter. Wash the macaroni, and boil it in the gravy and milk till quite tender. Drain it, put the macaroni into a very hot dish, and put by the fire. Beat the yolk of the egg with the cream and two tablespoonfuls of the liquor the macaroni was boiled in. Make this sufficiently hot to thicken, but do not allow it to boil, or it will be spoiled : pour it over the macaroni, and grate over the whole a little finely grated Parmesan cheese ; or the macaroni may be served as an accompaniment to minced beef, without the cheese ; or it may be taken alone with some good gravy in a tureen, served with it.

(9.) Stewed Eel.

One eel, half a pint of strong stock, two tablespoonfuls of cream, half a glass of port wine, thickening of flour, a little cayenne. Wash and skin the eel, cut it in pieces about two inches long, pepper and salt them and lay them in a stewpan, pour over the stock, and add the wine ; stew gently for twenty-five minutes or half an hour, lift the pieces carefully on a very hot dish and place it by the fire ; drain the gravy, stir into the cream sufficient flour to thicken it, mix with the gravy, boil for two minutes and add a little cayenne, pour over the eel, and serve.

(10). Custard Pudding.

Half a pint of milk, or a little more, two eggs. Warm the milk, whisk the eggs, yolk and white, pour the milk to them, stirring all the while. Butter a small basin that will exactly hold it, put in the custard, and tie a floured cloth over it, plunge into boiling water, turn it about for a few minutes. Boil it slowly for half an hour, turn it out and serve.

(11.) **Baked Custard.**

Half a pint of milk or a little more, two eggs. Warm the milk, whisk the eggs, yolk and white, pour the milk to them, stirring all the while, have ready a small tart dish lined at the edges with paste ready baked, pour the custard into the dish, grate a little nutmeg over the top, and bake in a very slow oven for half an hour.

(12.) **Essence of Beef.**

One pound of lean beef cut from the sirloin or rump, half-pint of cold water. Cut up the meat in small pieces, and place it in a covered saucepan by the side of the fire for four or five hours, then allow it to simmer gently for two hours, skim it well, and serve.

(13.) **Mutton Jelly.**

Six shanks of mutton, one and half quarts of water, pepper and salt to taste, half a pound of lean beef, a crust of bread toasted brown. Soak the shanks in water several hours, and scrub them well, put them and the beef and other ingredients into a saucepan with the water, and let them simmer very gently for five hours. Strain it, and when cold take off the fat; warm up as much as required when wanted.

(14.) **Nourishing Soup.**

Wash two ounces of best pearl sago well, then stew the sago in a pint of water till it is quite tender and very thick. Mix it with half a pint of good boiling cream and the yolks of two fresh eggs, mix the whole carefully with one quart of essence of beef, F. 12. The essence must be heated separately, and mixed while both mixtures are hot. A little of this may be warmed up at a time for use.

(15.) **Mutton Broth.**

One pound of the scrag end of neck of mutton, two pints of water, pepper and salt, half-pound of potatoes or some pearl barley. Put the mutton into a stewpan, pour over it the water, pepper and salt; when it boils skim it carefully, cover the pan, and let it simmer gently for an hour. Strain it, let it get cold, and then remove all the fat. When required for use add some pearl barley or potatoes in the following manner:—Boil the potatoes, mash them smoothly, see that no lumps remain, put the potatoes into a pan, and gradually add the mutton broth, stirring it till it is well mixed and smooth, let it simmer for five minutes and serve with fried bread.

(16.) Calf's-foot Broth.

One calf's-foot, three pints of water, one small lump of sugar, the yolk of one egg; stew the foot in water very gently till the liquor is reduced to half, remove the scum, set it in a basin till quite cold, and then take off every particle of fat. Warm up about half a pint, adding the butter and sugar, take it off the fire for a minute or two, then add the beaten yolk of the egg; keep stirring it over the fire till the mixture thickens, but do not let it boil or it will be spoiled.

(17.) Rabbit Soup.

Take a rabbit and soak it in warm water; when quite clean cut it in pieces and put it into a stewpan, and a teacupful of veal stock or broth, simmer slowly till done through, and then add one quart of water, and boil for an hour; take out the rabbit, pick the meat from the bones, covering it up to keep it white, put the bones back into the liquor and simmer two hours. Skim and strain and let it cool. Pound up the meat in a mortar with the yolks of two hard boiled eggs and the crumbs of a fresh roll previously soaked in milk; rub it through a tammy, and gradually add the strained liquor, and simmer for fifteen minutes. If liked thick, mix some arrowroot with half a pint of new milk, bring it to a boil, mix with the soup and serve. If preferred thin have ready some pearl-barley and vermicelli boiled in milk and add to the soup instead of the arrowroot. Serve with little squares of toast or fried bread.

(18.) Veal Soup.

Take a knuckle of veal, two cow heels, twelve peppercorns, one glass of sherry, two quarts of water. Put all these ingredients into an earthen jar, and stew six hours. Do not open it till cold. When wanted for use, skim off the fat and strain it; place on the fire as much as you require for use. Serve very hot.

(19.) Port Wine Jelly.

Take of port wine, 1 pint; isinglass, 1 oz.; sugar, 1 oz.; put the isinglass and sugar into ¼ pint of water; warm till all is dissolved, then add the wine, strain through muslin, and set to jelly. An excellent way of giving port wine.

(20.) Milk, with Rum, Whisky, or Brandy.

Put one tablespoonful of rum, brandy, or whisky, into half a pint of new milk, and mix well by pouring several times from one vessel to another. "Bilious" persons should heat the rum before adding it to the milk.—Dr. Dobell.

DIET TABLES.

In the treatment of disease the regulation of the diet always demands particular attention; it may be *made a powerful agent* in the restoration of health, while negligence or errors are often exceedingly disastrous. The writer has seen cases lost which, humanly speaking, might have been saved had the *nutrition* of the patient been properly regarded; in other words if he had had food in the proportion that his malady had medicine. It is quite beyond the scope of this little work to go into the principles which should govern the selection of food in disease. The utmost that our space will allow we have attempted. The reader will find the following tables assist him materially in determining a suitable diet for most cases.

LONDON FEVER HOSPITAL.

MEN.

Low Diet.

Bread 4 oz.; Milk ½ pint; Gruel 1 pint; Sugar ¼ oz.

Beef Tea Diet.

Beef Tea 1 pint; Milk 1 pint; Bread 4 oz.

Middle Diet.

Bread 10 oz.; Broth 1 pint; Milk 1 pint; Rice or Bread (for pudding) 2 oz.; Egg (for pudding) 1; Sugar (for pudding) ½ oz.

Fish Diet.

Bread 12 oz.; Fish (sole, haddock, cod, or brill, uncooked) 8 oz.; Potatoes 8 oz.; Cocoa 1 oz.; Sugar ½ oz.; Milk ⅛ pint.

Full Diet.

Bread 16 oz.; Meat (uncooked and without bone) 12 oz.; Potatoes 12 oz.; Cocoa 1 oz.; Sugar ½ oz. : Milk ¼ pint; Beer 1 pint.

Extras.

Beef Tea, Strong Beef Tea, and Eggs, as ordered; Arrowroot ½ oz.; Custard Pudding—1 Egg, ½ pint Milk, ½ oz. Sugar; Tea ¼ oz. per day; Sugar 1 oz. per day; Butter 1 oz. per day.

WOMEN.

Low Diet.

Bread 4 oz. ; Milk ½ pint ; Gruel 1 pint ; Sugar ¼ oz.

Beef Tea Diet.

Beef Tea 1 pint ; Milk 1 pint ; Bread 4 oz.

Middle Diet.

Bread 8 oz.; Broth 1 pint ; Milk 1 pint ; Rice or Bread (for pudding) 2 oz. ; Egg (for pudding) 1 ; Sugar (for pudding) ½ oz.

Fish Diet.

Bread 10 oz. ; Fish (sole, haddock, cod, or brill, uncooked) 8 oz. ; Potatoes, 8 oz.; Cocoa, 1 oz. ; Sugar ½ oz. ; Milk ½ pint.

Full Diet.

Bread 12 oz. ; Meat (uncooked and without bone) 10 oz. ; Potatoes 12 oz. ; Cocoa 1 oz. ; Sugar ½ oz. ; Milk ¼ pint ; Beer ½ pint.

Extras.

Beef Tea, Strong Beef Tea, Eggs, as ordered ; Arrowroot ½ oz. ; Custard Pudding—1 Egg, ½ pint Milk, ½ oz. Sugar; Tea ¼ oz. per day ; Sugar 1 oz. per day ; Butter 1 oz. per day.

HOSPITAL FOR DISEASES OF THE CHEST.

MEN.

Full Diet.

Breakfast.—Bread 12 oz. (*for the day*) ; Milk or Cocoa ½ pint.
Dinner.—Meat (cooked) 6 oz. ; Potatoes 8 oz. ; Porter 1 pint.
Supper.—Bread Pudding 8 oz., or Rice Pudding 8 oz., or Gruel or Corn-flour ½ pint (made with ¼ pint Milk), or Scotch Broth ½ pint.

Middle Diet.

Breakfast.—Bread 12 oz. (*for the day*) ; Milk or Cocoa ½ pint.
Dinner.—Meat (cooked) 4 oz. ; Potatoes 8 oz. ; Porter ½ pint.
Supper.—Bread Pudding 8 oz., or Rice Pudding 8 oz., or Gruel or Corn-flour ½ pint (made with ¼ pint Milk), or Scotch Broth ½ pint.

Milk Diet.

The day's allowance.—Bread 8 oz.; Milk 1¾ pint; Rice Pudding 8 oz., or Bread Pudding 8 oz.

Beef Tea Diet.

The day's allowance.—Bread 8 oz.; Milk 1½ pint; Beef Tea 1½ pint; 1 Egg.

WOMEN.

Full Diet.

Breakfast.—Bread 12 oz. (*for the day*); Milk or Cocoa ½ pint.
Dinner.—Meat (cooked) 4 oz.; Potatoes 8 oz.; Porter ½ pint.
Supper.—Bread Pudding 8 oz., or Rice Pudding 8 oz., or Gruel or Cornflour ½ pint (made with ¼ pint Milk), or Scotch Broth ½ pint.

Middle Diet.

Breakfast.—Bread 12 oz. (*for the day*); Milk or Cocoa ½ pint.
Dinner.—Meat (cooked) 3 oz.; Potatoes 8 oz.; Porter ½ pint.
Supper.—Bread Pudding 8 oz., or Rice Pudding 8 oz., or Gruel or Cornflour ½ pint (made with ¼ pint Milk), or Scotch Broth ½ pint.

Milk Diet.

The day's allowance.—Bread 8 oz.; Milk 1¾ pint; Rice Pudding 8 oz., or Bread Pudding 8 oz.

Beef Tea Diet.

The day's allowance.—Bread 8 oz.; Milk 1½ pint; Beef Tea 1½ pint; 1 Egg.

No extras to be supplied unless ordered in writing by the visiting Physician or Surgeon.

No extras allowed on full diet.

Every patient on being admitted into the Hospital to be placed on Beef Tea Diet until further orders.

KING'S COLLEGE HOSPITAL.

MEN.

DAILY ALLOWANCE.

Full Diet.

Breakfast.—Bread 6 oz. ; Milk ¼ pint.
Dinner.—Meat (cooked) 6 oz. ; Bread 6 oz. ; Potatoes ½ lb. ;
Porter 1 pint.
Supper.—Gruel 1 pint ; Milk ¼ pint.

Middle Diet.

Breakfast.—Bread 6 oz. ; Milk ¼ pint.
Dinner.—Meat (cooked) 4 oz. ; Bread 6 oz. ; Potatoes ½ lb. ;
Porter ½ pint.
Supper.—Gruel 1 pint ; Milk ¼ pint.

Milk Diet.

Breakfast.—Bread 4 oz. ; Milk ¼ pint.
Dinner.—Bread 4 oz. ; Rice Milk ½ pint (four days) ; Rice or
Bread Pudding ½ lb. (three days).
Supper.—Milk ½ pint.

CHILDREN'S DIETS (under ten years of age)—two-thirds of any
Diet ordered.

Roast Mutton—Monday and Thursday.
Boiled Mutton—Tuesday and Friday.
Stewed Mutton—Wednesday and Saturday.
On Sundays—Roast Beef.

WOMEN.

DAILY ALLOWANCE.

Full Diet.

Breakfast.—Bread 6 oz. ; Milk ¼ pint.
Dinner.—Meat (cooked) 4 oz. ; Bread 6 oz. ; Potatoes ½ lb. ;
Porter ½ pint.
Supper.—Gruel 1 pint ; Milk ¼ pint.

Middle Diet.

Breakfast.—Bread 6 oz. ; Milk ¼ pint.
Dinner.—Meat (cooked) 3 oz. ; Bread 6 oz. ; Potatoes ¼ lb. ;
Porter ½ pint.
Supper.—Gruel one pint ; Milk ¼ pint.

Milk Diet.

Breakfast.—Bread 4 oz. ; Milk ¼ pint.
Dinner.—Bread 4 oz. ; Rice Milk ½ pint (four days) ; Rice or Bread Pudding ½ lb. (three days).
Supper.—Milk ½ pint.

No extras (except Wine and Spirits) to be supplied by the Steward, unless authorized by the signature of the Visiting Physician or Surgeon.
No extras allowed on Full Diet.
In any Diet Rice or Bread Pudding may be substituted for Meat if desired.
No Patient on being admitted into the Hospital to be placed on Full Diet until ordered by the Visiting Physician or Surgeon.

THE MIDDLESEX HOSPITAL.

Convalescent Diet.

Daily.—12 oz. of Bread.
Breakfast.—½ pint Milk.
Dinner—MALE.—12 oz. of undressed Meat (leg and shoulder of Mutton only, except on Sundays, when the same quantity of Roast Sirloin and best Round of Beef is issued), roast and boiled alternately; ½ lb. of Potatoes.

FEMALE.—8 oz. of undressed Meat (leg and shoulder of Mutton only, except on Sundays, when the same quantity of Roast Sirloin and best Round of Beef is issued), roast and boiled alternately ; ½ lb. of Potatoes.

Supper.—1 pint of Gruel or 1 pint of Broth.

Half Convalescent Diet.

Daily.—12 oz. of Bread.
Breakfast.—½ pint of Milk.
Dinner.—4 oz. of undressed Meat (leg and shoulder of Mutton only, except on Sundays, when the same quantity of Roast Sirloin and Best Round of Beef is issued), roast and boiled alternately ; ½ lb. of Potatoes.
Supper.—1 pint of Gruel or 1 pint of Broth.

Pudding and Ordinary Diet.

Daily.—12 oz. of Bread.
Breakfast.—½ pint of Milk.
Dinner.—6 oz. of undressed Meat (Leg and Shoulder of Mutton only, except on Sundays, when the same quantity of Roast Sirloin and best Round of Beef is issued), roast and boiled alternately ; ¼ lb. of Potatoes ; 1 oz. Beef Suet, 2 oz. Flour for Pudding.
Supper.—1 pint of Gruel or 1 pint of Broth.

Ordinary Diet.

Daily.—12 oz. of Bread.
Breakfast.—½ pint of Milk.
Dinner.—6 oz. of undressed Meat (Leg and Shoulder of Mutton only), weighed with the bone before it is dressed—roast and boiled alternately; ½ Pound of Potatoes.
Supper.—1 pint of Gruel or 1 pint of Broth.

Half Ordinary Diet.

Daily.—12 oz. of Bread.
Breakfast.—½ pint of Milk.
Dinner.—3 oz. of undressed Meat (Leg and Shoulder of Mutton only), weighed with the bone before it is dressed—roast and boiled alternately ; ¼ lb. of Potatoes.
Supper.—1 pint of Gruel or 1 pint of Broth.

Mutton Broth Diet.

Daily.—12 oz. of Bread.
Breakfast.—½ pint of Milk.
Dinner.—8 oz. of undressed Meat (Neck of Mutton only), weighed with the bone before it is dressed—served in 1 pint of Broth with Barley.
Supper.—1 pint of Gruel.

Fish Diet.

Daily.—12 oz. of Bread.
Breakfast.—½ pint of Milk.
Dinner.—8 oz. of Fish (whiting, sole, haddock, cod, plaice, or brill); ½ pound of Potatoes.
Supper.—1 pint of Gruel.

Milk Diet.

Daily.—12 oz. of Bread.
Breakfast.—½ pint of Milk.
Dinner.—*Alternate days.*—2 oz. Rice Pudding, ½ Egg, ½ oz.
Sugar ; 1½ oz. Sago Pudding, ½ Egg, ½ oz. of Sugar ; Bread Pudding, 1½ Egg, ¾ oz. Sugar. *Extra.*—Custard ½ oz.
Supper.—½ pint of Milk.

Simple Diet.

Daily.—12 oz. of Bread.
Breakfast.—½ pint of Milk.
Dinner.—1 pint of Gruel.
Supper.—½ pint Milk.

Extras.

For supper, Meat when cooked, 3 oz. ; Chops, ½ lb. each when trimmed ; ordinary Beef Tea, ½ lb. of Clod and Sticking of Beef, without bone, to a pint ; Strong Beef Tea, 1 lb. of Clod and Sticking of Beef, without bone, to a pint ; Broth, without Meat, ¼ lb. of Neck of Mutton with bone, to a pint—this Broth is made with that for the Patients on Mutton Broth Diet ; Steaks—Rump Steaks, ¼ lb., without bone ; Tripe ; Chicken ; Oysters ; Greens ; Eggs ; Arrowroot ; Sago ; Jellies ; Porter ; Wine ; Spirits.

Every Patient admitted into the Hospital is placed upon Simple Diet, until a Diet is otherwise ordered.

No Extras are to be placed on the Diet Table, nor provided by the House Steward, other than those specified above.

Diets ordered by the Physicians and Surgeons are to be continued until changed by subsequent orders.

Extras are allowed for one day only, unless the Physician or Surgeon write the word DAILY.

On Sundays those Patients on Meat Diet have Roast Beef.

Breakfast is served at 7, Dinner at 12, Tea at 4, and Supper at 7 o'clock.

LONDON HOSPITAL.

Fancy Diet.

MEN AND WOMEN.

Per day.—12 oz. Bread ; 8 oz. Potatoes ; 1 pint Porter.
Breakfast.—*Alternate Days.*—Gruel and 1 Egg ; Gruel and Slice of Meat.

Dinner.—Sunday, Roast Mutton and Rice Pudding ; Monday, Fish and Batter Pudding ; Tuesday, Rabbit and Light Pudding ; Wednesday, Roast Mutton and Rice Pudding ; Thursday, Fish and Batter Pudding ; Friday, Rabbit and Light Pudding ; Saturday, Roast Mutton and Rice Pudding ; Sunday, Fish and Batter Pudding ; Monday, Rabbit and Light Pudding ; Tuesday, Roast Mutton and Rice Pudding ; Wednesday, Fish and Batter Pudding ; Thursday, Roast Mutton and Rice Pudding ; Friday, Fish and Batter Pudding ; Saturday, Rabbit and Light Pudding.

Supper.—1 pint Broth.

CHILDREN (under seven years of age).

12 oz. Bread, ½ pint Milk, daily ; 2 oz. Meat and 8 oz. Potatoes five times a week; and Rice Pudding twice a week.

Extras.

Mutton Chops ; Beef Steaks ; Fish ; Beef Tea ; Strong Broth ; Rice Pudding—1 oz. of Rice, ½ oz. of Sugar, ½ pint of Milk, ¼ oz. of Butter ; Light Pudding—½ pint of Milk, ½ oz. of Sugar, 2 Eggs ; Batter Pudding—½ pint of Milk, 2 oz. of Flour, ½ oz. of Sugar, ½ oz. of Suet, 1 Egg ; Eggs ; Bread; Green Vegetables ; Watercresses ; Wine ; Spirit ; Porter.

Extras discontinued unless order renewed by the Physician or Surgeon at each visit.

The ordinary Diets of this Hospital do not call for special notice.

SPECIAL DIETS, ETC.

DIETETIC TREATMENT OF SPASMODIC ASTHMA.

The diet must be regularly weighed out, and adhered to with the greatest strictness, the hours of meals being most rigidly fixed as follows:—

Breakfast at 8 a.m.—To consist of half a pint of green tea or coffee, with a little cream, and two ounces of dry, stale bread.

Dinner at 1 p.m.—To consist of two ounces of fresh beef or mutton, without fat or skin, and two ounces of dry stale bread or well-boiled rice ; three hours *after* dinner (not sooner) half a pint of weak brandy and water, or whiskey and water, or dry sherry and water, may be taken, or toast and water *ad libitum.*

Supper at 7 p.m.—To consist of two ounces of meat as before, with two ounces of dry stale bread.

The patient is not to be allowed to drink any fluid whatever within one hour *before* his dinner or supper, and not until three hours after either of these meals. At other times he is not limited as to drinks, otherwise than that all malt liquors are to be prohibited. *Soda* or *Seltzer* water may be indulged in at other times when thirsty.

With this dietetic treatment sedatives are to be given as follows :—

Three grains of the Extract of Conium are to be taken four times a-day ; namely, at the hours of seven, twelve, five, and ten, the dose to be gradually increased to five grains four times a-day. To each of these pills a *fourth* of a grain of the Extract of Indian Hemp may be added, which may be gradually increased to one grain in each dose.

MR. PRIDHAM, *Brit. Med. Journal,* June 9 to Dec. 29, 1860.

DIETETIC TREATMENT OF APOPLEXY.

The diet of the patient should be low, till all apprehension of a relapse is passed, and limited to milk, boiled vegetables, light puddings, and fish. At no subsequent period ought he to indulge in a full animal diet, or to drink undiluted wines. At the same time, too lowering a regimen is to be avoided, as thereby the irritability of the system and the heart's action generally is increased. All the causes of the disease already fully referred to should be avoided, counteracted, or overcome.

Dr. AITKEN, *"Science and Practice of Medicine,"* vol. ii., page 505.

DIETARY IN CASES OF SLOW DIGESTION.

Breakfast (8 a.m.).

Bread (stale), 4 oz. { Mutton chop or other meat (cooked), free from fat and skin, 3 oz. } Tea, or warm milk and water and sugar, or other beverage, ¾ pint.

Luncheon (1 p.m.).

Bread (stale), 2 oz. { No solids, such as meat or cheese. } Liquid, ¼ pint.

Dinner (5 or 6 p.m.).

Bread (stale), 3 oz. Potatoes and other vegetables, 4 oz. { Meat (cooked), free from skin and fat, 4 oz. } Liquid not more than ½ pint.

Tea or Supper (not sooner than three hours after dinner).

Bread (stale), 2 oz. { No solids such as meat or cheese. } Tea or weak brandy and water, or sherry and water, or toast and water, to the extent of ½ pint.

Dr. LEARED.

DIETETIC TREATMENT IN EPILEPSY.

In the adult the diet should be light, and the patient should live temperately. He should live *by rule*. He should rise early and take regular exercise in the open air, keeping his head cool and his feet warm. The diet of an infant so affected should be, if possible, its mother's milk, with or without arrowroot. If above three or four years of age, its diet should consist entirely of farinaceous or of other light vegetable food.

Dr. AITKEN, "*Science and Practice of Medicine,*" vol. ii., page 543.

REGIMEN AND DIET IN LOSS OF NERVE POWER, ETC.

REGIMEN AND DIET IN LOSS OF NERVE POWER, ETC., always of importance in the treatment of every case, is especially so in those diseases in which Phosphorus or its combinations are indicated.

Attention should, in the first place, be directed to the general health of the patient, condition of the digestive organs, and state of the bowels; where these are disordered suitable correctives should be prescribed before commencing as well as during the course of

Phosphorus. Constipation, so frequently persistently present, is to be relieved; and where there is torpidity of liver, mild mercurials, when not contra-indicated, or podophyllin in alterative doses, should be administered, combined with gentle aperients. The following numbers, 63, 65, 119, 141, 163, are suggested as being well adapted to these purposes. They will effectually prevent the clogging of the liver of which some writers speak.

It is to be borne in mind that Phosphorus partakes rather of the nature of food than of medicine; its office is to supply nutrition of a special character, and its operation is *promoted* by a judicious, and *delayed*, if not destroyed, by an injudicious management; it is *important* therefore that the patient should submit to a *prescribed regimen*, enjoining abstinence from, or extreme moderation in, the use of alcoholic drinks and tobacco, the avoidance of over-feeding, mental fatigue, strong muscular exercise, and excesses of every kind, that directly or indirectly fatigue, not to say exhaust, the forces and waste the tissues which the phosphorus is administered to support and nourish. The adoption of early hours and the daily use of the bath (cold or tepid, according to the condition of the patient; the cold bath, either plunge, douche, or sponge, always to be preferred when admissible), and exercise in the open air, are all absolutely necessary. *The Turkish Bath* is a valuable help in the treatment of many disorders, and it may always be safely employed where there is no reason to suspect organic lesion. It is especially useful in nervous exhaustion from overwork. I know of no remedy which so quickly relieves *brain fatigue.* The diet must be simple and nutritious, and good of its kind; fresh meat, mutton or beef, plainly dressed and not over-cooked, fish and fresh vegetables. Eggs, recently laid, uncooked, are at once easy of digestion and highly nutritious, in some cases as many as half a dozen may be taken in twenty-four hours. In cases where wine is advisable Burgundy is to be preferred; tea and coffee, unless they be very moderately taken, had better be altogether avoided. The *kind*, as well as the *quantity*, of the food must however be determined by the digestive power of the patient, and this is found to be frequently so impaired as only to admit of milk and and easily assimilated farinaceous substances, such as lentil meal, corn flour, macaroni, etc.

So important is the determination of the diet in the treatment of nervous disease—often dependent more or less entirely upon defective nutrition—that it should in every case be made the subject of a *special prescription*, in which the kind and quantity of the food should be definitively stated for the guidance of the patient.

DIET FOR DIABETES.

The important principle to be regarded is the exclusion of starch and sugar, and articles containing them ; all else is of little consequence. Meat, poultry, fish, game, green vegetables, cress, celery, lettuce, spinach, and the like ; butter, cheese, eggs, are all admissible. Van Abbott's and Banthoron's gluten bread and biscuits, and Blatchley's bran biscuits are all very valuable substitutes for wheaten bread.

Pepsine, Pancreatine, and Dilute Hydrochloric Acid, are very useful *digestants* in diabetes.

Dr. Pavy says diabetic patients

May drink—

Dry Sherry, Claret, Sauterne, Soda Water, Burton Bitter Beer.

Must NOT drink—

Milk (except sparingly), Sweet Ales, Porter, Stout, Cider, Sparkling Wines.

K

INDEX TO FORMULÆ.

Mixtures and Glycecols are arranged Alphabetically.

Pilulæ.

INDEX OF DISEASES AND REMEDIES.

Adynamic Fevers.—Alcohol; Ammonia; Mist. Ammon. Acet.; Mist. Ammon. Acet. Co.; Mist. Cinchonæ Amm. et Chloroformi; F. 176, 186.

After-Pains.—Morphia Granules; Opium Granules; F. 176, 177; Tr. Opii c. Chloroformi.; Liq. Opii Sed.; Syr. Chloral.

Ague.—See FEVER INTERMITTENT.

Albuminuria.—*Acute Stage* (Purgation). F. 51, 60, 126. Mist. Magnes. Sulph. et Rosæ; Mist. Ammon. Acet. Co. *Chronic Stage*—Ol. Morrhuæ; Mist. Acid. Nitrohydrochlor. c. Ferro; Mist. Tonici; F. 35, 36, 132.

Amenorrhœa.—F. 45, 68, 69, 82, 87, 105, 106, 107, 196; Ergotine Granules; Pil. Aloes et Ferri, B.P.; Pil. Ferri Iodidi, B.P.; Pil. Aloes et Myrrhæ, B.P.; Pil. Assafœtidæ Co., B.P.; Mist. Acid. Nitrohydrochlor. c. Ferro; Decoct. Aloes Co.

Anæmia.—The preparations of Iron, especially Ferrum Redactum, in the form of Glycecol or in combination with Phosphorus, F. 193 and 195; F. 74, 76; Elixir Pyrophosphate Iron; Elixir Bark and Iron.

Angina Pectoris.—Nitrite Amyle; Chloroform; Counter-irritation.

Animal Poisoning.—Ammonia; Alcohol.

Anthrax.—See CARBUNCLE.

Anus, Fissures of.—Tannin; Carbolic Acid; Borax (Suppositories and Glycerina).

Anus, Pruritis.—Chloroform; Opium (Suppositories); Glycerina Belladonnæ; Local Anæsthetics.

Aphonia, Catarrhal.—Glycecol Aluminis, Zingiberis, Acid. Benzoic., Acid. Tannic., Acid. Tannic. et Capsici; Turpentine Inhalation; Counter-irritation.

Appetite, Loss of.—Mist. Acid. Sulph. Arom.; Tr. Cardam Co. c. Quinâ; Elixir Bark.

Aphthæ and Aphthous Ulcerations.—Glycecol Potassæ Chloratis, Sodæ Sulphitis, Boracis, Sulphurous Acid.

Apoplexy.—Croton Oil; F. 56, 126; Croton Oil Enemata; Calomel.

Apoplexy, Congestive.—Venesection.

Asphyxia.—Cold Affusion; Electricity; Artificial Respiration.

Asthma.—F. 33, 89, 94, 152, 153, 158. Glycecol Belladonnæ, Lobeliæ; Tinct. Veratri Viridis; Mist. Ammoniaci, Ipecac. et Lobeliæ; Fuming Inhalations. See SPECIAL DIET.

Baldness.—Linimentum Crinale.

Bilious Attacks.—F. 60, 61, 62, 119, 141. Mist. Alkalina. Arom. c. Rheo; Mist. Mag. Sulph. c. Rosâ; Mist. Alkalina (Soda) c. Calumbâ.

Bites, Venomous.—Suction; Ligature; Ammonia; Alcohol.

Brain Fatigue.—Phosphorus.

Breath, Fœtor of the.—Glycecol Acid. Carbolic., Potassæ Chloratis, Acid. Tannic.

Bronchitis, Acute.—F. 98, 99, 100, 111, 133, 134, 135; Mist. Ammon. Acet. (Simp. et Comp.).

Bronchitis, Chronic.—F. 15, 20, 102, 183; Mist. Ammoniaci, Ipecac. et Lobeliæ; Mist. Ammon. Senegæ; Glycecol Lobeliæ, Ipecac. et Morphiæ, Ammonii Chloridi, Althææ, and Cubebæ; Turpentine Stupes.

Burns and Scalds.—Pigmentum Collodio c. Ol. Ricini.

Bursæ.—Pigmentum Iodi.

Cancer of Rectum.—Iodoform.

Cancer and Cancerous Affections.—F. 21, 71.

Cancrum Oris.—Glycecol Potassæ Chloratis; Nitric Acid and Bark Mixture.

Carbuncle.—Liquor Potassæ; Bark and Ammonia Mixture; Pil. Quinæ, B.P.; F. 70, 84, 125, 196, 191.

Catarrh.—F. 98 to 100, 102, 103, 165, 167; Glycecol Camphoræ, Potassæ Chloratis, Doveri, Jacobi, Ammon. Mur.; Mist. Ammon. Acet.; Mist. Ammon. Acet. Co.

Chancres, etc.—F. 7, 29, 176, 177. Iodoform.

Chilblains, Chapped Hands, etc.—Glycerinum Zinci Co.

Chlorosis.—Iron, see ANÆMIA.

Cholera, Asiatic.—F. 2, 40, 138; Tinct. Opii Etherea.; Tinct. Opii et Chloroformi.; Mist. Acid. Sulph. Arom. See DIARRIŒA, PREMONITORY.

Diarrhœa, Premonitory.—Glycecol Confect. Opii; Kino Comp.; F. 2, 138; Mist. Astringens c. Hæmatoxyli; Chlorodyne.

Diarrhœa, Chronic.—F. 25, 37, 38. Mist. Acid. Nitrohydrochlor.

Diarrhœa of Children.—F. 39; Mist. Astringens; Pulv. Astringens.; Glycecol Confect. Aromat., Krameriæ, Catechu, Gummi Rubri, Hæmatoxyli.

Diphtheria.—Lactic Acid (local); Mist. Tonici.

Dropsy.—F. 9, 22, 52, 101, 104; Mist. Ammon. Acet.; Mist. Diuretica; Mist. Senegæ Ammon.

Dysmenorrhœa.—F. 71, 89, 177; Belladonna Granules; Acetate Ammonia (Mixture).

Dyspepsia.—F. 25, 45, 47, 48, 49, 50, 57, 58, 73, 83, 156, 172, 180. Mist. Alkalina Aromat.; Mist. Alkalina c. Gentianâ; Mist. Alkalina c. Calumbâ; Mist. Alkalina Arom. c. Rheo.; Elixir Bismuth.; Elixir Pepsin.; Glycecol Bismuthi, Pepsin., Sodæ Rhei et Zingib., Sodæ Carb., and Lupulinæ.

Dysentery.—Ipecacuanha, Əj. to 3j. doses; Calomel and Opium, F. 2; Opium Granules; Bismuth; Glycerine.

Eczema.—F. 24, 50, 121, 127, 146; Mist. Potassii Iodidi Co.; Mist. Mag. Sulph. Acid.; Mist. Alba; Mist. Acid. Phosph. c. Ferro, with Arsenic; Locally Pig. Zinci Co.

Epilepsy.—F. 18, 26, 79, 185; Potassii Bromid.; Oxide Zinc Pills; Mist. Tonici.

Epistaxis (Bleeding at the Nose).—Mist. Rosæ with Gallic Acid; Cold, Plugging.

Erysipelas.—Mist. Cinchon. Ammon.; Mist. Acid. Phosph. c. Ferro; Mist. Alba; Podophyllin.

Fevers, Remittent.—Pulv. Salinæ Effervescens.; Mist. Salinæ; Mist. Diaphoreticæ; Emetics(Ipecac.); Mist. Mag. Sulph. Acidi; Quinine; Mist. Cinchon. Acid.; Cinchonine; Mist. Chirettæ.

Fevers, Intermittent (Ague, etc.).—Quinine Preparations; Arsenic; Cinchonine; Chloride Ammonium; Piperin; Mist. Alba; Mist. Salinæ; Mist. Diaphoretica.

Fever, Scarlet.—*Emetics;* Ipecac., Aconite (Granules); Belladonna (Granules); Chlorate Potash (Glycecol and F. 103); Alcohol; Quinine, F. 125; Mist. Cinch. Ammon.; Mist. Cinchon. Acid.; Mist. Acid. Nitrohydrochlor.; Potus Chlori; Potus Potassæ Chloratis; Potus Acidi; Mist. Acid. Phosph. c. Ferro.

Fever, Typhus.—*Emetic*—Ipecac.; Potus Acid. Hydrochlor.; Mist. Acid. Nitrohydrochlor. *Stimulants*—Alcohol; Opium; Belladonna. *During Convalescence*—Mist. Cinchon. Acid.; Mist. Acid. Phosph. c. Ferro; Mist. Chiratæ Co.; Quinine and Iron; F. 144, 70, 81, 74.

Fever, Typhoid.—*Emetic*—Ipecac.; Potus Acid.; Mist. Acid. Nitrohydrochlor; Mist. Acid. Sulph. Arom.; Opium; Acid. Gallic., F. 25, 37, 38. *Convalescent Stage*—Quinine, F. 144, 70, 81, 74, 75; Mist. Cinchonæ Acidi; Mist. Chiratæ Co.; Mist. Tonici.

Gastralgia (Heartburn).—F. 49, 50, 109; Arsenic, F. 24; Mist. Alkalina c. Gentianâ; Mist. Alkalina c. Calumbâ; Glycecol Carbonis, Sodæ Carb., Sodæ Rhei et Zingib.

Gastritis.—F. 25, 37, 46. Counter-irritation.

Gastrodynia.—Elixir Bismuth; Glycecol Bismuthi.

Gleet.—F. 108, 162. Mist. Acid. Phosph. c. Ferro; Catheterism and Injections.

Gonorrhœa.—F. 108, 164, 176, 177. Mist. Copaibæ Co.; Mist. Rosæ Aperiens; Injections.

Gout.—F. 11, 22, 23, 30, 114, 115, 116, 159, 187; Mist. Magnes. Sulph. Alkalina; Tr. Colchici Etherea; Tr. Guaiaci Etherea; Glycecol Colchici, Lithiæ, Pigmentum Belladonnæ, locally applied.

Hæmoptysis.—F. 36, 130; Opium; Pil. Plumbi c. Opio, B.P.; Tannic and Gallic Acids, Persulphate Iron, and Acetate Lead, applied locally in an atomized form; Mist. Acid. Sulph. Arom.; Mist. Rosæ.; Ice.

Hæmorrhages.—F. 36, 130. Mist. Mag. Sulph. c. Rosâ; Mist. Acid. Sulph. Aromat.; Liquor Ergotæ; Oil of Turpentine; Tannic Acid; Ice.

Hæmorrhoids, Piles.—Mistura Alba; F. 10, 60, 128; Glycecol Confect. Piperis; Tannin; Iodoform (Suppositories).

Headache, Dyspeptic or Sick.—Guarana, F. 46; Mist. Alkalina Arom.

Headache, Nervous.—Ammonia; Valerianate Ammonia; F. 185, 182.

Headache, Congestive.—F. 22, 62. Mist. Rosæ Aperiens; Mustard Pediluvium.

Headache, Rheumatic or Gouty.—Mist. Alba; F. 22; Mist. Alkalina; Potash.

L.

Headache, Intermittent.—See NEURALGIA.

Hoarseness.—Glycecols Acid. Benzoic., Cubebæ, and Althææ; Linctus Tussi; Spray Solutions; Steam Inhalations.

Hypochondria.—See MELANCHOLIA.

Hysteria.—F. 18, 26, 32, 34, 59, 68, 78, 88, 105, 145, 182, 185. Bromide Potassium (Mixture); Phosphorus.

Impotence.—Phosphorus, F. 189 to 200; Nux Vomica.

Jaundice.—F. 53, 163; Podophyllin, F. 53, 119; Calomel; Ipecac.; Mist. Alkalina .(Potash) c. Gentianâ; Mist. Acid. Nitrohydrochlor.; Hot Baths.

Laryngitis.—Glycerine; Turpentine Stupes.

Lepra.—F. 24, 121, 127. Mist. Potassii Iodidi Co.; Mist. Hydrargyri Co. et Sarsæ.

Leucorrhœa.—Tannic Acid (locally as Lotion or Pessary); Iron Alum; Permanganate Potash Lotion; Mist. Acid. Phosph. c. Ferro; F. 81, 87, 162.

Liver, Torpid.—F. 10, 53, 119; Mist. Acid. Nitrohydrochlor.; Mist. Chiratæ Co.

Liver, Congested,—F. 22, 60; Mist. Alba.

Lumbago.—Mist. Alba; Tr. Colchici Ether.; Iodide Potassium, F. 11, 30, 155; Hot Baths; Turkish Bath; *Liniments—* Aconiti, Pot. Iodidi, Terebinth.

Malarial Poisoning.—Quinine, F. 80, 178; Cinchonine, F. 173; Arsenic, F. 80; Mist. Cinchoniæ Co.

Menorrhagia.—F. 36, 132, 35. Iron Alum; Turpentine; Acetate Lead; Pessaria Ferri Perchloridi; Mist. Acid. Sulph. Arom.; Liquor Ergotæ; Mist. Rosæ Aper.

Melancholia.—F. 189 to 200, 77, 85, 88. Alcohol; Alteratives.

Nervous Affections.—See DEBILITY, MELANCHOLIA, NEURALGIA, HYSTERIA.

Neuralgia.—F. 21, 32, 71, 78, 79, 84, 122, 184, Phosphorus 189 to 200. Syr. Chloral Hydrate; Chloroform; Croton Chloral.

Paralysis.—Aperients; F. 51, 22, 126, 174, 56; Counter-irritation; Electricity; Strychnia, F. 85, 77; Iron, F. 106, 81, 168, 144, 179. Mist. Tonici; Syr. Ferri Phosph.; Phosphorus; Nux Vomica.

Peritonitis (Inflammation of Bowels).—F. 1 to 6; Opium Granules; Aconite Granules; Tinct. Veratri Virid.; Counter-irritation; Turpentine Stupes.

Pericarditis (Rheumatic).—F. 1 to 6; Blisters; Turpentine Stupes. See REMEDIES FOR RHEUMATISM.

Pertussis (Whooping-Cough).—F. 18, 31, 93, 136; Glycecol Bromide Ammonium, Bromide Ammonium and Belladonna, Sulphate Zinc and Belladonna, Lobeliæ.

Phthisis.—*Perspirations*—F. 35, 36, 130, 160, 167; Mist. Cinchon. Acid. *Cough*—F. 111; Inhalations; Glycecol Aconite, Opium, Morphia, Pruni. Virg.; Linctus pro Tussi; Glycecol Morph. et Ipecac. *Dyspnœa*—F. 152, 153. *Diarrhœa*—F. 38, 37; Syr. Rhatany; Mist. Astringens c. Hæmatoxyli. *Tonics*—Cod Liver Oil; Mist. Tonici; Mist. Cinchon. Acid.; Mist. Cinchon. Ammon.; Mist. Chiratæ Co.; Phosphorus. *Counter-irritation*—Pigmentum Iodi; Blistering Colloid; Turpentine Stupes.

Pleurisy.—F. 1 to 6, 90, 98, 99, 100. Mist. Diaphoretic; Counter Irritants; Turpentine Stupes.

Pleurodynia.—F. 49, 79, 156. Bismuth Glycecol.

Pneumonia, Acute.—Tinct. Verat. Vir.; Aconite Granules; Alcohol; Belladonna Granules; Tartar Emetic; F. 135, 134, 165, 99, Mist. Salina; Mist. Diaphoretica; Turpentine Stupes and Poultices.

Psoriasis.—F. 17, 24, 121, 124, 127.

Pyrosis.—F. 49, 156. Elixir Bismuthi; Mist. Ammon. Effervescens with Prussic Acid; Mist. Acid. Sulph. Arom.; Glycecol Bismuthi, Tannin, Oxalate Cerium.

Rheumatic Gout.—Alteratives; Tr. Guaiaci Ether.; Lithia; Turkish Bath. See REMEDIES FOR RHEUMATISM.

Rheumatism, Acute.—F. 98 to 100, 103, 114 to 116, 159, 181, 187; Mist. Alba; Mist. Diaphoretica; Mist. Alkalina (Potash); Tr. Colchici Ether.; Baths.

Rheumatism, Chronic.—F. 11 to 13, 30, 82, 155, 175, 187; Mist. Potassii Iodidi Co.; Tr. Guaiaci Ether.; Baths (Turkish).

Rubeola (Measles).—F. 99, 102. Mist. Salinæ; Potus Potassæ Bitart.; Syrupus Ipecacuanhæ; Glycecol Ipecac.

Scabies.—Sulphur; Sulphide Calcium.

Scarlatina.—See SCARLET FEVER.

Sciatica.—F. 56, 126, 155, 11, 30, 13. Phosphorus, 193 to 200; Blisters; Lin. Aconiti.

APPENDIX.

Heim's Pills.

℞ Pulv. Digitalis, gr. x.; Pulv. Ipecac., gr. v.; Pulv. Opii, gr. v.; Ext. Helenii, q.s. ft. pil. xx.

Dose—One pill three times a day.

Use—As an antipyretic in phthisis these pills give excellent results. Fever being a most active source of exhaustion, remedies which have the power of controlling it, lessening its intensity and duration, are exceedingly valuable. Dr. Niemeyer, in his *Practical Medicine*, speaks highly of this combination and the following.

Heim's Pills with Quinine.

℞ Quiniæ Sulph., gr. xx.; Pulv. Digitalis, gr. x.; Pulv. Ipecac., gr. v.; Pulv. Opii, gr. v.; Ext. Helenii, q.s. ft. pil. xx.

Dose—One three times a day.

Digitalis and Quinine have a well-merited reputation as a means of arresting abnormal calorification and reducing animal heat. This preparation is especially appropriate in cases of phthisis, when fever of a periodic type, marked by chills and evening exacerbations, is present. The effect of these pills, like other preparations containing Digitalis, should be watched. They may be suspended when a distinct reduction of temperature, and the frequency of the pulse is apparent, and resumed as occasion may require.

Dr. Martini's Anti-Hysterical Pills.

℞ Auri et Sodii Chloridi, gr. v.; Pulv. Tragacanth et Sacchari, q.s. ft. pil. xl.

Dose—One pill to be taken an hour after dinner and supper. Afterwards two pills at these hours, and gradually increase the dose up to eight pills daily.

This remedy is mentioned also by Dr. Niemeyer as a *nervine* of great efficacy in hysteria, and he has used it with signal effect in many cases in which there was no indication for the local treatment of uterine affections.

Ricord's Pills.

℞ Hydrargri Protoiodidi, Lactucarii, āā ℨjss. ; Ext. Opii Aquosæ, gr. ix. ; Ext. Guaiaci Aquosæ, ℨj. Ft. pil. xxxvj.

Dose—One twice or thrice a day.

Use—In constitutional syphilis.

These pills are found to produce in many subjects severe pain in the bowels. This, however, is more the fault of the dose, which is excessive, than the remedy, which is found to give excellent results in smaller doses. Formula 14 is a safer preparation and rarely disagrees.

Blaud's Pills.

℞ Ferri Sulph. Pur., Potassæ Carb. Pur., āā ℨss. ; Pulv. Tragacanth, q.s. ft. pil. xcvj.

(The original recipe orders 48).

Use—Said to be a specific in chlorosis.

"For twenty years," says Dr. Niemeyer, "I have used *Blaud's* pills in chlorosis, and have witnessed such brilliant results that I have found no opportunity to experiment with any other preparation."

Dose—Three pills, increased to four or five if well borne, thrice daily.

Croton Choral Pills.

℞ Croton Chloral, gr. xxiv. ; Pulv. Tragacanth, gr. vj. ; Ext. Gentianæ, q.s. ft. pil. xij. =gr. ij. in each pill.

A new remedy, for which we are indebted to Liebreich. Employed to produce anæsthesia of the fifth nerve, it has been found exceedingly useful in trigeminal neuralgia. Sometimes employed as a soporific, in doses of five to ten grains. Dr. Yeo recommends it in irritative night cough in phthisis, in acute neuralgia, in doses of from 2 to 5 grains every hour or the smaller dose every half hour until 15 grs. have been taken.

Dose—From 1 to 3 pills.

For full description of this remedy see an excellent article by Dr. Yeo in the *Lancet*, July 31*st*, 1874.

Sulphide Calcium Pills.

℞ Calcii Sulph., gr. ij. ; Pulv. Tragacanth et Ext. Gentianæ, q.s. ft. pil. xx.

Dose—One pill every two or three hours.

Use—"The sulphides appear often to arrest suppuration. In inflammation threatening to end in suppuration they reduce the inflammation and avert the formation of pus. In *boils* and *carbuncles* they yield excellent results. A tenth of a grain of Sulphide of Calcium given every two or three hours, generally prevents the formation of fresh boils, while it lessens the inflammation and reduces the area of the existing boils and quickly liquefies the core, so that its separation is much more speedy, thus considerably curtailing the course of the boil."—RINGER's *Therapeutics*.

See the *Lancet*, Feb. 21, 1874.

Cinchonine Pills.

℞ Cinchoninæ Hydrochlor., gr. xxiv.; Pulv. Capsici, gr. iv.; Ext. Gentianæ, q.s. ft. pil. xij.

Dose—One to two pills once or twice daily.

It is a matter of clinical experience that Capsicum increases the action of the Cinchona Alkaloids. It certainly aids their rapid diffusion when given in a pilular form. The present high price of Quinine gives the Cinchonines additional value as its substitute, and they certainly deserve to be more largely employed. "A commission having in 1866 been appointed in the Madras Presidency to examine the antiperiodic powers of Cinchonine, Cinchonidine and Quinidine, supplies of these alkaloids were placed at the disposal of medical officers at 'noted malarious stations,' and were tested by 1145 cases of paroxysmal fevers of all types. * * * The main conclusion which the members of the commission have derived from the data before them is, 'that these alkaloids, hitherto little valued in medicine, are scarcely, if at all, inferior as therapeutical agents to Quinine.' As a general rule it appears that those experiments were the most successful in which *medium* doses (gr. ij. to gr. v.) were administered in a single dose daily, the cases recovering more expeditiously than when larger or smaller quantities were employed."—*Pharmacopœia of India*, 1868.

For combination with Iron and Nux Vomica see F. 173.

Bromide Mercury Pills.

(*Hospital for Women.*)

℞ Hydrargyri Bromidi, gr. vj. ; Confect. Rosæ, q. s. ft. pil. xij.

Dose—One or two pills.

This must not be confounded with the Bibromide of Mercury, a preparation stated to be analogous in its action to Corrosive Sublimate.

Aconite Granules.

The virtues of Aconite are daily becoming more appreciated. That it is a remedy of very great value is acknowledged by all therapeutists, and a convenient mode of administering it accurately in small doses is a desideratum. The officinal extract (prepared from the leaves) gives very uncertain results, a disadvantage which is not noticed when prepared from the root by alcoholic exhaustion. It is found that this alcoholic extract, when finely divided with sugar of milk, and made into granules, is very uniform in its operation.

See note, page 66.

Oxalate of Cerium Pills.

℞. Cerii Oxalatis, gr. xxiv.; Sacchari Lactis, gr. xij.; Ext. Glycyrrhizæ, q.s. ft. pil. xij.

Uses—In irritable dyspepsia, attended with gastrodynia, pyrosis, vomiting, especially the vomiting of pregnancy. In chorea, epilepsy, and other allied convulsive affections.

Dose—One or two pills. Often found to succeed in cases where Nitrate of Silver and Bismuth have failed.

Black Snake Root Pills.

℞. Ext. Acteæ Racemosæ Alcoholic., gr. xxxvj.; Pulv. Tragacanth et Glycyrrhizæ, q.s. ft. pil. xij.

This medicine is highly esteemed in America, and the root from which it is prepared, although not officinal, has been used in England.

Its properties are said to be *alterative, antispasmodic, diaphoretic, expectorant, resolvent, emmenagogue, parturient,* etc.

Dose—One pill three times a day.

This remedy was thought well of by Dr. James Simpson, of Edinburgh, and Dr. Neligan thinks it deserves trial.

Iridin or Irisin Pills (derived from *Iris Versicolor*).

℞. Iridin, gr. xij.; Sacchari Lactis, gr. xij.; Ext. Glycyrrhizæ, q.s. ft. pil. xij.

Uses—In *scrofula, syphilis, gonorrhœa, dropsy,* rheumatism, glandular swelling, eruptions of skin, and affections of liver or spleen.

Dose—One or two pills three times a day.

"Irisin is justly," says Dr. Glover Coe, "esteemed as one of our most valuable remedies. It is eminently resolvent, and exercises a marked influence over the whole glandular system, quickening the activity of the secreting apparatus, and promoting depuration."

Butler & Tanner, The Selwood Printing Works, Frome, and London.